DATE DUE

AG 5 '99			
AP 2 3 '01			
MY 1 4 '01			
JE 5 '01			
DE 1 8 '03			

DEMCO 38-296

Ergonomic Design of Material Handling Systems

Ergonomic Design of Material Handling Systems

Karl H. E. Kroemer, Dr. Ing.

Professor and Director, Industrial Ergonomics Laboratory
Human Factors Engineering Center
Department of Industrial and Systems Engineering
Virginia Tech
Blacksburg, Virginia

LEWIS PUBLISHERS

Boca Raton   New York

Library of Congress Cataloging-in-Publication Data

Kroemer, K. H. E., 1933–
    Ergonomic design of material handling systems / Karl H. E.
    Kroemer.
        p.   cm.
    Includes index.
    ISBN 1-56670-224-0
    1. Lifting and carrying.   2. Materials handling.   3. Human
engineering.   I. Title.
T55.3.L5K76   1997
621.8′6—dc21   97-12784
                                                                CIP

© 1997 by CRC Press LLC
Lewis Publishers is an imprint of CRC Press LLC

No claim to original U.S. Government works
International Standard Book Number 1-56670-224-0
Library of Congress Card Number 97-12784
Printed in the United States of America 1 2 3 4 5 6 7 8 9 0
Printed on acid-free paper

About the Author

Karl H. E. ("Eb") Kroemer's interests are in engineering applications of ergonomic knowledge, especially in anthropometry, biomechanics, and work physiology, and in related research. Since 1981, he has been a professor at Virginia Tech (VPI&SU) where he directs the Industrial Ergonomics Laboratory. He is also a director of the ERI, the Ergonomics Research Institute, Inc., a consulting firm in Radford, Virginia. He can be reached easily by e-mail under karlk@vt.edu, by telephone (540) 231-5677 or 639-0514 and via P.O. Box 1309, Radford, VA 24143-3019.

His academic degrees are in mechanical engineering from the Technical University Hannover in Germany. He worked 7 years as a research engineer at the German Max-Planck Institute for Work Physiology and 3 years as director of the Ergonomics Division of the B.A.U. (the "German NIOSH"). In the U.S.A., he worked 7 years as research industrial engineer in the U.S. Air Force's Human Engineering Division, and he spent 3 years as professor of ergonomics and industrial engineering at Wayne State University in Detroit. He has also served as U.N. ergonomics expert in Romania and India. He has consulted extensively with industry, government agencies, and universities in America, Europe, and Asia and as an ergonomics expert in law cases.

Consulting, teaching, and research are Dr. Kroemer's fortes. He conducts workshops, seminars, and courses, ranging from the basic "Introduction to Ergonomics" to special topics in human engineering, for example on avoiding cumulative trauma disorders, setup and use of computer workstations, ergonomic management of material handling, matching work requirements with human strength and endeavor and, in general, on the design of tasks, processes, and equipment to fit human capabilities.

Preface

Ergonomics *is the study of human characteristics for the appropriate design of living and work environments.*[1]

The term "ergonomics" was devised in 1950 in England from the Greek words "ergon," relating to work and strength, and "nomos," indicating law or rule. In the United States and in Canada, the term "human factors" is often used.

Human Factors/Ergonomics adapts the manmade world to people, focusing on the human as the most important component of our technical systems.

Ergonomics/Human Factors spans the whole range from basic research to engineering and managerial applications. They have major roots in the biological sciences (anthropology, physiology, and medicine), in the behavioral sciences (particularly experimental and engineering psychology), in the engineering sciences (particularly the industrial, mechanical, and computer disciplines), and in their interactions, such as occupational medicine, industrial hygiene, bioengineering, or biodynamics.

Ergonomics/Human Factors utilize the approaches and techniques of their parent disciplines, combine these with new methodologies and tools (e.g., biomechanics or computer-aided systems modeling) to optimize concepts, designs, development, manufacturing, testing, management, and human participation in manned systems.

Ergonomics/Human Factors is:

- human-centered
- trans-disciplinary
- application-oriented.

The goal of Human Factors/Ergonomics is "humanization" of working and living conditions. This goal can be symbolized by the two "E's," Ease and Efficiency for which technological systems and all their elements should be designed. This requires knowledge of the characteristics of the people involved, particularly of their dimensions, capabili-

---

[1] Kroemer, K.H.E., Kroemer, H.B., and Kroemer-Elbert, K.E. (1994). *Ergonomics: How to Design for Ease and Efficiency.* Englewood Cliffs, NJ: Prentice Hall.

ties, and limitations, and the conscientious engineering and managerial application of this knowledge. Purposeful consideration of human needs and abilities contributes essentially to successful, efficient, and safe performance of technological systems, and to satisfying work and leisure.

HANDLING LOADS

If we move a sizeable object from point A to point B using our hands, we speak of "load handling." This may mean that the mass is brought vertically to a different height, horizontally to a different distance, or held in place and possibly turned about a pivot. We usually divide material handling into groups of activities: lifting and lowering, pushing and pulling, holding, carrying, and turning. We create these groupings for the convenience of separate analyses and ergonomic design, but in reality these actions are often combined. In fact, some load handling is part of other work, such as inspection, assembly, cleaning, polishing, sorting, positioning, or placing.

There is a curious use of terminology. Obviously, "material handling" originally meant moving a load by hand, but the expression has also come to be applied when machinery of some kind does the job. To make clear that a human does the task, the compound term "manual material handling" has been created: this is not only a word monster but also a perfect tautology because "manual" means "by hand." (That is similar to the "foot pedal.") In this text, the term *handling* always indicates that human hands move the object.

ERGONOMIC LOAD HANDLING

Ergonomics can be applied both in the original design of material handling systems and in the modification of existing ones. Ergonomics supplies information on people as supervisors, equipment operators, or load handlers. Application of ergonomic knowledge ensures prudent use of human capabilities and abilities, and safeguards people from overexertion and undue strain.

Ergonomics can be used in two major strategies:

Fitting the person to the job. This means selection of individuals for their ability to perform certain tasks, and training these persons to perform their task better and more safely.

Fitting the job to the person. Here, the task, equipment, and work organization are adjusted to fit human capabilities, limitations, and preferences.

Especially in material handling, both approaches can be used at the same time to supplement each other. However, *fitting the job to the person has highest priority.*

Ergonomics overlaps and intertwines with traditional engineering and management, in fact uniting them. Ergonomics provides the information for matching the work to the operator, and for finding suitable operators and training them. Hence, ergonomics achieves EASE for people and EFFICIENCY in material handling.

Karl H. E. Kroemer
Blacksburg, VA
April 1997

Table of Contents

Key 1   FACILITY LAYOUT

In the real world, one encounters either one of two situations. In the first case, a facility exists, it must be used as found, and the building and its interior layout cannot be changed significantly. One must make the best of what there is. In the second case, one can plan and design a new facility and its details according to the process at hand, to suit the product and the production.

The opportunity to "do it right at the drawing board" is most desirable, allowing the closest approximation to the optimal solution. Striving for the best possible solution is also the purpose for modification of a given facility. Therefore, the ideal case will be used here to guide even when only modifications may be possible.

It is the purpose of facility layout, or facility improvement, to select the most economical, efficient, and safest design of building, department, and workstation. Of course, specific details depend on the overall process.

A facility with well laid out material flow has short and few transportation lines. Transportation is always costly in terms of space, machinery, and energy; it does not add value to the object being moved; and it is full of hazards to people. In many existing facilities, reduction and simplification of material movement can lower the expense of material transport considerably, which often amounts to 30 to 75 percent of total operating cost. Of course, even more important from a human point of view is the possibility of reducing the risks of overexertion and injury to workers by redesigning, improving, or eliminating transport lines.

> **Ergonomic Check 1-1. FEW TRANSPORTS**
>
> A facility that is well laid out has
>
> • few and short transportation lines.

Of course it is much more efficient to design a facility — from the very beginning of the planning — for ergonomically best transport rather than trying to improve a design that is faulty. For this reason it is very important to include an ergonomist in the team that is planning a new facility.

> **Ergonomic Check #1-2. ERGONOMIC PLANNING**
>
> To design a facility that is laid out for minimal transport and the least human load handling
>
> • have an ergonomist on the design team.

Instead of simply moving an object from one workstation to the next, interrupting the flow for performing work on the piece, one can consider performing work on the object while it moves. Automobile assembly lines are early and well-known examples of the "continuous flow" solution. This may work well if tools and instruments, parts and components all move along with the work piece, but may be impossible in other cases, for example, if finely controlled precision work is necessary.

PROCESS OR PRODUCT LAYOUT

There are two major design strategies: *process layout* and *product layout*. In the first case, all machines or processes of the same type are grouped together, such as all heat treating in one room, all production machines in another section, and all assembly work in a different division. Figures 1-1 and 1-2 are examples of process layout.

There is a major advantage to this process layout design: Quite different products or parts may flow through the same

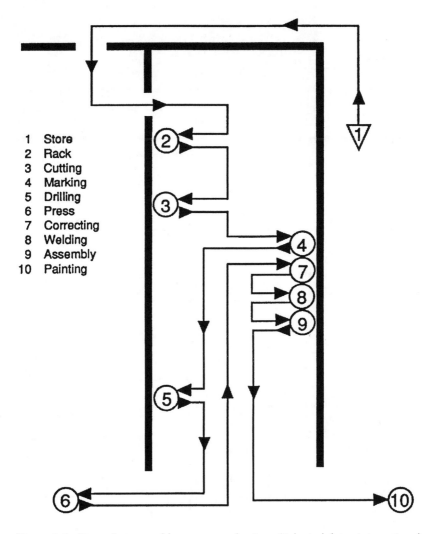

1 Store
2 Rack
3 Cutting
4 Marking
5 Drilling
6 Press
7 Correcting
8 Welding
9 Assembly
10 Painting

Figure 1-1 Flow diagram of bus seat production. (Adapted from International Labour Office, 1974.)

workstations, keeping machines busy. But much floor space is needed, and there are no fixed flow paths. Process layout requires a relatively large amount of material handling. It is worthwhile to study Figures 1-1 and 1-2 carefully to determine what improvements should be made in each case to improve the conditions depicted.

In contrast, in product layout all machines, processes, and activities needed for the work on the same product are grouped together. This results in short throughput lines, and relatively

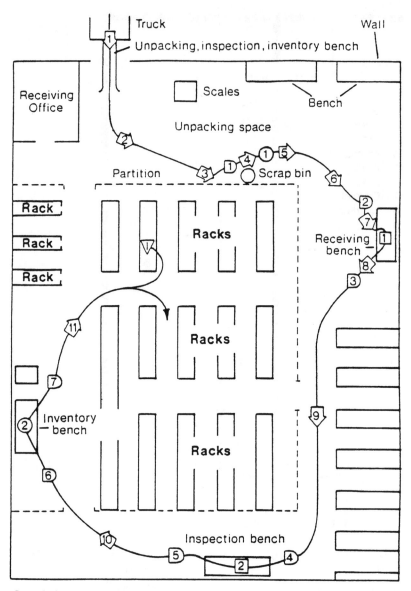

Receiving - present setup

Figure 1-2 Flow diagram of receiving, inspection, inventorying, and storage: original setup. (Adapted from International Labour Office, 1974.)

little floor space is needed. However, the layout suits only the specific product, and breakdown of any single machine or of special transport equipment may stop everything. Altogether, product layout is advantageous for material handling because routes of material flow can be predetermined and planned well in advance.

FLOW CHARTS AND DIAGRAMS

It is rather easy to describe events and activities with simple sketches, symbols, and words.

The *flow diagram* is a picture or sketch of activities and events. It indicates their sequence, and where they take place. Figure 1-1 shows the production of bus seats.

The *flow chart* is a listing or table of the same activities and events. It indicates their duration and provides detailed information on related facts or conditions.

Table 1-1 describes the steps taken in making flow diagrams and charts.

Figure 1-2 depicts the flow of an item from arrival to storage in a receiving facility, and Figure 1-3 presents more details. Figures 1-4 and 1-5 provide descriptions of an improved setup. Figure 1-6 is a blank sample of a flow chart meant to analyze load handling.

Of course, modern technology allows much more flexible recording, and then analysis, of material flow by using computer templates and software instead of the traditional paper-and-pencil forms. The flow charts, for example, can be modified to reflect special conditions in specialized industries, or to point out specific concerns, such as repetitive manipulations. (See the section on cumulative trauma in Key 2.)

Table 1-1 Making Flow Chart and Diagram

1. SELECT THE ACTIVITY TO BE STUDIED:
   Be sure that you are really studying what you need to know.

2. CHOOSE THE SUBJECT (operator or material) TO FOLLOW:
   Pick the person or material depending upon which goes through the entire process on which you are working. Stick with it. Every detail must be recorded on that subject. (WATCH IT: Be careful not to confuse the subject — either person or material — being followed. If a person carries an object, puts it down, and then goes to get a truck, and we are following the person, we do not show a delay symbol for the object. It may be resting, but the worker is not.)

3. DETERMINE STARTING AND ENDING POINTS:
   This is to make sure you cover everything you want to cover, but no more.

4. WRITE A BRIEF DESCRIPTION OF EACH DETAIL (in the chart):
   Step by step, no matter how short or temporary, describe every operation, every inspection, every transfer, and every delay. Write it on the chart as you see it done.

5. APPLY THE SYMBOLS:
   The description determines the symbol. (You may create your own symbols, if needed!) Draw connecting lines between proper symbols.

◯	OPERATION:	When something is *intentionally* being changed, created, or added, use a large O to show that an "operation" or action is done. (Picking up an object, tightening a bolt, writing a letter.)
▢	INSPECTION:	When something is checked or verified but not changed, use a square to denote an "inspection." (Checking a requisition, measuring a machine part.)
⇨	TRANSPORT:	When something is moved from one place to another, use an arrow to show "transportation." (Taking a part out of the supply bin, carrying packaging material to the dump.)
▽	STORAGE:	When something is put intentionally ("formally") in a place for some time (as in a file or in a storage area), use the triangle to indicate "storage."
D	DELAY:	When something remains in one place awaiting action, use the large D to show "delay." (A carton waiting to be lifted.)

   Number each series of symbols as they are used.

6. ENTER FACTS (in the chart):
   Enter distance, time, weight, size of object, frequency of occurrence, number of people involved, or other facts describing the activity.

7. DESCRIBE HAZARDS:
   Mark whether material might fall, or has sharp edges on which one could be cut, whether there are pinch points, if material is "handled," etc. Rate the degree of hazard as related to other jobs (this is obviously subjective).

Table 1-1 Making Flow Chart and Diagram (Continued)

8. SUMMARIZE:
Add up all facts and hazards and put the totals in the "summary" block.

9. COMMENTS:
After the flow chart is completed, study all details of the job in order to determine what improvements can be made. Jot down problems observed, and/or possible improvements.

10. CONTROL ACTION:
Question each activity in order to establish suitable control actions. The following scheme helps.
WHY is this activity necessary? An unsatisfactory answer here may lead to the elimination of this element of the job or perhaps of the complete job.
WHAT is being done? What is the purpose of doing it? We want to know if this activity achieves what it is supposed to. Often, one can simplify the activity.
WHERE is the activity being done? Where is the best place to do the details? *WHY* should it be done there? Be sure that if the detail is necessary it is done in the right place. You will often find that the activity can be combined with others.
WHEN is the activity done? Why should it be done by now? Be sure that (if the activity is necessary) it is done at the right time.
WHO does it? Who should do it? Why is it done this way? Can we make the activity easier to do and safer for both personnel and equipment? Or should PPE (personal protective equipment) be worn? Use of PPE indicates that you were not able to alleviate the hazard: it is still there.

11. PROPOSE A BETTER WAY:
The preceding steps should have made quite clear where problems lie, what could be done, and what should be done. Propose a better way. Document it on a flow chart form. Summarize the details of your proposal in the box on the chart. Let the differences shown there be your best arguments.

SUMMARY

	PRESENT		PROPOSED		DIFFERENCE	
	No.	Time	No.	Time	No.	Time
○ Operations	2	8				
⇧ Transportations	11	26				
☐ Inspections	2	35				
D Delays	7	85				
▽ Storages	1	2				
Distance travelled	61 m		m			
Number of hazards — High	0					
Medium	8					
Low	8					

FLOW PROCESS CHART

No. 1   Page 1 of 2

JOB: Receive, check, inspect, inventorize and storage of parts received in cartons.

☐ OPERATOR OR ☒ MATERIAL

CHART BEGINS 9:15 a.m.
CHART ENDS 11:31 a.m.
CHARTED BY KHEK   Date

METHOD	PRESENT OR PROPOSED (Describe in Detail Each Shop)	FACTS — Distance m	Time min	Weight kg	Size m.m.m	Freq/shift	# of people	HAZARDS — Falling mat	Sharp edges	Pinch points	Hazard mat	Manual handling	Overall rating	Highmat/flow	COMMENTS (Identify Important Aspects, Particularly Hazard Potentials)	CONTROL ACTION (Eliminate / Combine / Simplify / Personal Protection etc.)
1	Carton lifted from truck; placed on inclined roller conveyor	2	2	75	0.7 0.7 0.5	4 0 0	2	✓				✓	L		Back injury possible	Change
2	Slid on conveyor	6	1	·	·	·	2	✓					L			Combine
3	Stacked on floor	6	4	·	·	·	2			✓			L			Eliminate
4	Await unpacking	-	32										-			Eliminate
5	Unstacked	1	2	·	·	·	2	✓		✓			M			Eliminate
6	Lid removed, shipment papers removed	-	3	·	·	·	2		✓				M			Change
7	Place on hand truck	1	1	·	·	·	2			✓			M		Back injury possible	Eliminate

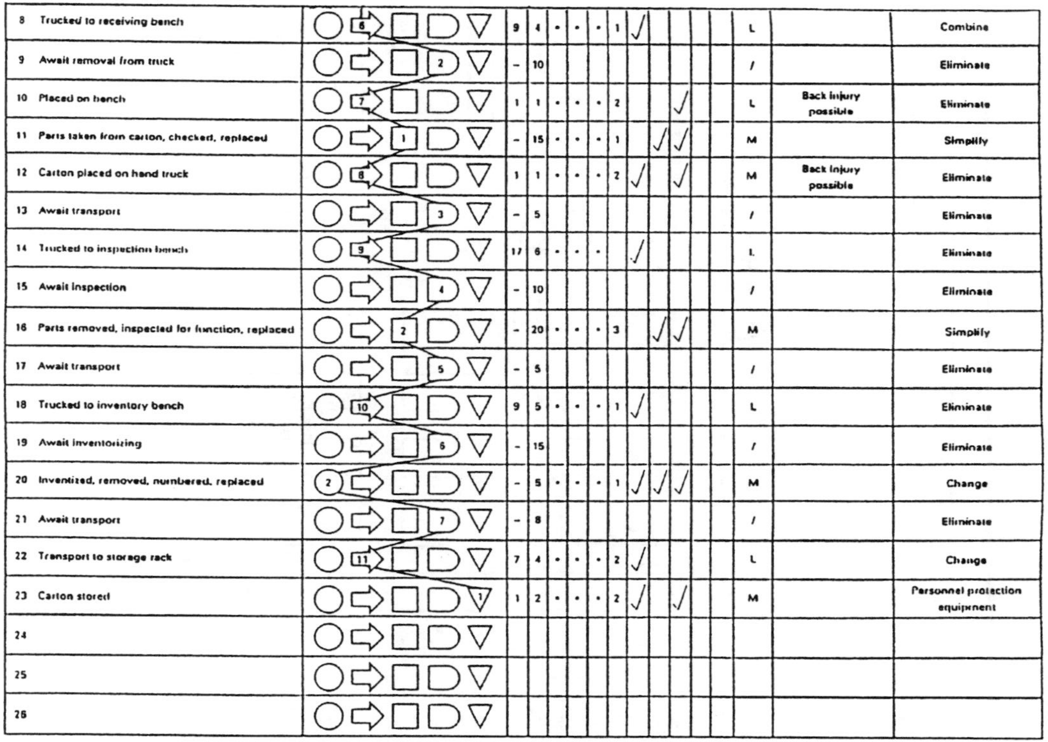

Figure 1-3 Flow chart of the setup shown in Figure 1-2.

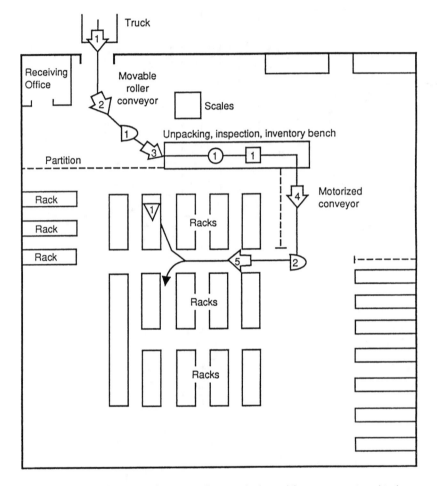

Figure 1-4 Flow diagram of improved setup. (Adapted from International Labour Office, 1974.)

FLOW PROCESS CHART

No. 1
Page 1 of 2

JOB Receive, check, inspect, inventorize, and storage of parts.

☐ OPERATOR or ☒ MATERIAL
CHART BEGINS --
CHART ENDS --
CHARTED BY KHEK DATE 1/30/98

SUMMARY

	PRESENT		PROPOSED		DIFFERENCE	
	No.	Time	No.	Time	No.	Time
○ OPERATIONS	2	8	2	6	-	(1B)
⇨ TRANSPORTATIONS	11	26	5	7	6	19
☐ INSPECTIONS	2	35	1	25	1	35
D DELAYS	7	85	2	25	5	60
▽ STORAGES	1	2	1	2	-	-
DISTANCE TRAVELLED	61 m		30 m		31 m	
NUMBER OF HAZARDS	High 0		0		0	
	Medium 8		2		6	
	Low 8		4		4	

METHOD PRESENT or PROPOSED	ACTIVITY					FACTS						HAZARDS						COMMENTS	CONTROL ACTION
DESCRIBE IN DETAIL EACH STEP	OPERATION	TRANSPORT	INSPECTION	DELAY	STORAGE	DISTANCE, m	TIME, min.	WEIGHT, kg	SIZE, m·m·m	FREQ./SHIFT	# OF PEOPLE	FALLING MAT.	SHARP EDGES	PINCH POINTS	HAZARD. MAT.	MANUAL HANDLING	OVERALL RATING High/Med./Low	IDENTIFY IMPORTANT ASPECTS, PARTICULARLY HAZARD POTENTIALS	ELIMINATE COMBINE SIMPLIFY PERSONAL PROTECTION EQUIPMENT
1 Lift from truck on roller conveyor	○	⇨	☐	D	▽	2	2	75.7	1.5	400	2	✓					L		
2 Slide on conveyor to bench	○	⇨	☐	D	▽	121		"	"	"	1	✓					L		
3 Wait on conveyor	○	⇨	☐	D	▽	-	22	"	"	"	-								
4 Slide on conveyor to combined inspection/inventory	○	⇨	☐	D	▽	2	1	"	"	"	1	✓					L		
5 Unpack part, inspect, inventorize, place on conveyor	○	⇨	☐	D	▽	6	26	"	"	"	2	✓	✓				M		
6 Move by motorized conveyor to pickup	○	⇨	☐	D	▽	5	1	"	"	"	-								
7 Await pickup	○	⇨	☐	D	▽	-	3	"	"	"	-	✓							
8 Pick up by stacker truck	○	⇨	☐	D	▽	4	2	"	"	"	1	✓	✓				L		
9 Store in rack	○	⇨	☐	D	▽	1	2	"	"	"	1	✓					M		

Figure 1-5 Flow chart of improved setup shown in Figure 1-2.

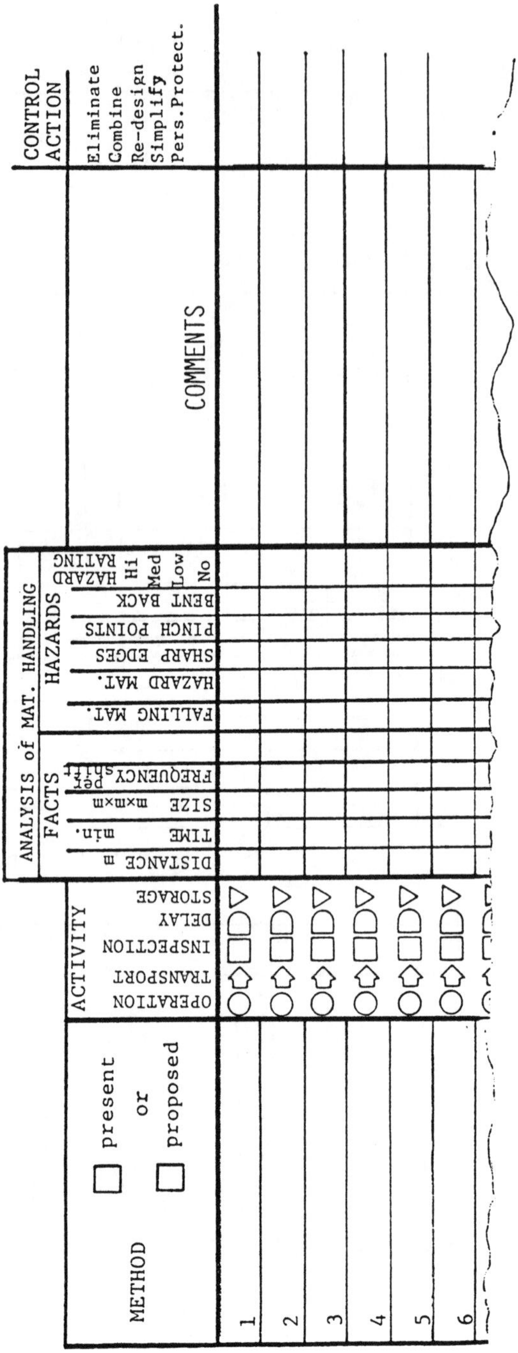

Figure 1-6 Sample of a flow chart specifically designed for analysis of load handling.

Key 2   JOB DESIGN

This is perhaps the most important decision point in the planning and design of a new facility, or in the improvement and redesign of an existing plant. Which material movement jobs should be assigned to machinery? What activities need to be performed by people?

Task allocation primarily falls into three categories:

- Automation of the material movement, with no people directly involved.
- Mechanization of the material movement, with people as controllers and operators.
- Manual material movement, with persons doing the job at the risk of injury and overexertion.

Machines are much superior to people in speed, power, and their ability to perform routine, repetitive, and precise actions. Furthermore, people function well as supervisors and controllers; they can deal with unexpected events, can make decisions even when only incomplete information is available, and are able to function under overload conditions. This speaks for as much automation, or mechanization, as is technologically and economically feasible. Hence, a strong plea can be made for mechanization and automation, although this may mean the elimination of jobs that were unduly hazardous in the first place.

The Material Handling Institute has been promoting "principles of material movement." Several of these can help in task allocation and job design. These are listed in Table 2-1.

Ergonomic Check 2-1. AUTOMATE

Automation and mechanization help to avoid injuries and overexertion

• No people — no injuries.

These principles include some recommendations made in Key 1 regarding planning and automation. The "simplification principle" is part of classic industrial methods engineering of the middle 1900s, but now has acquired special meaning: with the recent recognition of the dangers associated with repetitive motions, specialization and simplification at the job should not be carried to the extent that a person must perform the same body motions and exertions thousands of times, over and over.

Table 2-1   Principles of Material Movement

Safety Principle Provide suitable methods and equipment for safe handling.	Make people operators and controllers of equipment, instead of having them do manual material movement.
Planning Principle Plan all material movement activities to obtain maximal efficiency and safety.	Use graphic and tabular aids; also, assembly chart, operation process chart, travel chart, activity relationship chart, critical-path diagram, etc., and other tools of the industrial engineer.
Automation/Mechanization Principle Design personal risk out of the process by having machinery do the job.	For example: automated stackers; powered conveyors; robots.
Simplification Principle Simplify material flow by eliminating, reducing, or combining activities.	For example: use one-piece die cast aluminum housing instead of milling and drilling several parts and then welding them together.

Source: Adapted from Kulwiec, R. A., Ed., (1985) *Material Handling Handbook,* 2nd Ed., New York: Wiley.

JOB DESIGN TO AVOID CUMULATIVE TRAUMA DISORDERS

Cumulative trauma disorders (CTDs) occur when an activity is repeated so often that it overloads the body parts involved.

Such overuse disorders have been described[2,3] in the literature under various names since the early 1700s — washer woman sprain, telegraphist cramp, pianist myalgia, trigger finger, typist wrist, and the repetitive strain injury (RSI) of keyboard operators. Table 2-2 lists those conditions that are most often associated with cumulative trauma disorders (CTDs). Of course, this list is neither complete nor exclusive. New occupational activities occur, and several activities may be part of the same job.

Table 2-2   Common Repetitive Strain Injuries

Disorder	Causes and symptoms
Carpal tunnel syndrome (writer's cramp, neuritis, median neuritis, partial thenar atrophy)	The result of compression of the median nerve in the carpal tunnel of the wrist. This tunnel is an opening under the carpal ligament on the palmar side of the carpal bones. Through this tunnel pass the median nerve, the finger flexor tendons, and blood vessels. Swelling of the tendon sheaths reduces the size of the opening of the tunnel and pinches the median nerve, and possibly blood vessels. The tunnel opening is also reduced if the wrist is flexed or extended, or ulnarly or radially pivoted. Force and frequency of finger movements, and hammering, are typical stressors.
Cubital tunnel syndrome	Results from compression of the ulnar nerve distal to the notch of the elbow; often related to putting the arm on a hard surface, especially an edge.
DeQuervain's syndrome (or disease)	A special case of tendosynovitis (see below) that occurs in the abductor and extensor tendons of the thumb where they share a common sheath. This condition often results from combined forceful gripping and hand twisting, as in wringing cloths.
Epicondylitis ("tennis elbow")	Tendons attaching to the epicondyle (the lateral protrusion at the distal end of the humerus bone) become irritated from overuse or strain. This condition os often the result of impacting or jerky throwing motions, repeated supination and pronation of the forearm, and forceful wrist extension movements. The condition is well-known among tennis players, pitchers, bowlers, and people who do hammering. (A similar irritation of the tendon attachments on the inside of the elbow is called medial epicondylitis, also know as "golfer's elbow.") This is often associated with jerking motions, forearm twisting, and forceful wrist extension.

[2] Kroemer, K.H.E., Kroemer, H.B., and Kroemer-Elbert, K.E. (1994). *Ergonomics: How to Design for Ease and Efficiency*. Englewood Cliffs, NJ: Prentice Hall.
[3] Kuorinka, I. and Forcier, L. (Eds.) (1995). *Work Related Musculoskeletal Disorders — A Reference Book for Prevention*. London: Taylor & Francis.

Table 2-2   Common Repetitive Strain Injuries (Continued)

Disorder	Causes and symptoms
Ganglion	A tendon sheath swelling that is filled with synovial fluid, or a cystic tumor at the tendon sheath or a joint membrane. The affected area swells up and causes a bump under the skin, often on the dorsal or radial side of the wrist. (Since it was in the past occasionally smashed by striking with a bible or heavy book, it wa also called a "bible bump.") This condition is often associated with forceful hand closure and forearm twisting.
Neck tension syndrome	An irritation of the levator scapulae and trapezius group of muscles of the neck, commonly occurring after repeated or sustained overhead work.
Pronator teres syndrome	Result of compression of the median nerve where it passes through the two heads of the pronator teres muscle in the forearm. This often occurs after strenuous flexion of elbow and wrist.
Shoulder tendinitis (rotator cuff syndrome or tendinitis, supraspinatus tendinitis, subacromial bursitis, subdeltoid bursitis, partial tear of the rotator cuff)	This is a shoulder disorder located at the rotator cuff. The cuff consists of four tendons that fuse over the shoulder joint, where they pronate and supinate the arm and help to abduct it. The rotator cuff tendons must pass through a small bony passage between the humerus and the acromion, with a bursa as cushion. Irritation and swelling of the tendon or of the bursa are often caused by carrying a load on the shoulder or by continuous muscle and tendon effort to keep the arm elevated.
Tendinitis (tendonitis)	An inflammation of a tendon. The tendon becomes thickened, bumpy, and irregular in its surface. Tendon fibers may be frayed or torn apart. In tendons without sheaths, such as within elbow and shoulder, the injured area may calcify. Often associated with repeated tension, motion, bending, being in contact with a hard surface, vibration.
Tendosynovitis (tenosynovitis, tendovaginitis)	This disorder occurs to tendons that are inside synovial sheaths. The sheath produces excessive synovial fluid which accumulates, and the sheath swells. Movement of the tendon within the sheath is impeded and painful. If tendon surfaces become irritated, rough, and bumpy, and the inflamed sheath presses progressively onto the tendon, the condition is called stenosing tendosynovitis. ("DeQuervain's syndrome" (see previous page) is a special case occurring in the thumb, while the "Trigger finger" (see following page) condition occurs in flexors of the fingers.) The disorder often occurs with forceful hand closure and twisting motions.

Table 2-2   Common Repetitive Strain Injuries (Continued)

Disorder	Causes and symptoms
Thoracic outlet syndrome (neurovascular compression syndrome, cervicobrachial disorder, brachial plexus neuritis, costoclavicular syndrome, hyper-abduction syndrome)	A disorder resulting from compression of nerves and blood vessels between clavicle and first and second ribs at the brachial plexus. If this neurovascular bundle is compressed by the pectoralis minor muscle, blood flow to and from the arm is reduced. This ischemic condition makes the arm numb and limits muscular activities. Often associated with overhead work, keeping the upper arm elevated, carrying a load with an extended arm.
Trigger finger	This is a special case of tendosynovitis (see there) where the tendon becomes nearly locked, so that its forced movement is not smooth but occurs in a snapping jerking manner. This is a special case of stenosing tendosynovitis crepitans, a condition usually found with finger flexors. It is often associated with using hand tools that have sharp, hard edges or whose handles are too far apart for the user's hand so that the end segments of the fingers are flexed while the middle segments are straight.
Ulnar artery aneurism	Weakening of a section of the wall of the ulnar artery in the Guyon tunnel in the wrist. The resulting bubble presses on the ulnar nerve. Often found in assembly workers.
Ulnar nerve entrapment (Guyon tunnel syndrome)	Results from the entrapment of the ulnar nerve as it passes through the Guyon tunnel in the wrist (similar to the carpal tunnel syndrome). It can occur from prolonged flexion and extension of the wrist and repeated pressure on the hypothenar eminence of the palm.
White finger ("dead finger," Raynaud's syndrome)	Stems from insufficient blood supply bringing about noticeable blanching. Finger turns cold, numb, and tingly, and sensation and control of finger movement may be lost. The condition is due to closure of arterial vessels caused by vasospasms triggered by vibrations. A common cause is continued forceful gripping of vibrating tools, particularly in a cold environment.

Adapted from Kroemer, K. H. E. (1992). Avoiding Cumulative Trauma Disorders in Shop and Office. *J. Ind. Hyg. Assoc.* 53(9), 596–604. With permission.

CTD-PRONE ACTIVITIES AND POSTURES

The major job factors in CTDs are rapid, often-repeated movements, forceful exertions and movements, static muscle loading including maintenance of odd posture, and vibrations, especially in a cold environment. Their negative effects are likely to be aggravated if several components are combined, such as numerous repetitions with high force exerted in awkward posture.

High *repetitiveness* has been defined as a cycle time of less than 30 seconds, or as more than 50% of the cycle time spent performing the same fundamental motion. High *force* exerted with the hand, e.g., more than 45 N, may be a causative factor by itself. Posture may be highly important. For example, a "dropped" or "elevated" wrist, or a wrist bent to either side, reduces the available cross-section of the carpal tunnel in the wrist (through which pass the flexor tendons of thumb and fingers, blood vessels, and the median nerve), and hence generates a condition that may contribute to the development of a disease called carpal tunnel syndrome.

A strongly maintained isometric contraction of muscles needed to keep the body or its parts in position, or to hold a hand tool, to carry an object, is often associated with a CTD condition. Also, inward or outward twisting of the forearm with a bent wrist, a strong deviation of the wrist from the neutral position, and the pinch grip can be stressful.

Seven Sins. There are seven conditions that specifically need to be avoided in job design:

1. Job activities with many repetitions.
2. Work that requires prolonged or repetitive exertion of more than about one third of the operator's static muscular strength available for that activity.
3. Putting body segments in extreme positions
4. Work that makes a person maintain the same body posture for long periods of time.
5. Pressure from tools or work equipment on tissues (skin, muscles, tendons), nerves, or blood vessels.
6. Work in which a tool vibrates all or part of the body.
7. Exposure of working body segments to cold, including air flow from pneumatic tools.

Ergonomic Check 2-2. TELL-TALE SIGNS OF CTDs

Overuse disorders are most likely with

- rapid and often repeated actions
- repetitive exertion of finger, hand, or arm forces
- pounding and jerking with the hands and arms
- contorted body joints
- polished-by-use sections of workplace, clothing; use of custom-made padding
- blurred outlines of the body owing to vibration
- the feeling of cold and the hissing sound of fast-flowing air.

The features of CTDs are various, variable, and often confusing. The onset of their symptoms can be gradual or sudden. Three stages have been defined.

Stage 1 is characterized by local aches and tiredness during the working hours, which usually abate overnight and with days away from work. There is usually no reduction in work performance. This condition may persist for weeks or months and is reversible.

Stage 2 has symptoms of tenderness, swelling, numbness, weakness, and pain that start early in the workshift and do not abate overnight. Sleep may be disturbed and the capacity to perform the repetitive work is often reduced. This condition usually persists over months.

Stage 3 is characterized by symptoms that persist at rest and during the night; pain occurs even with non-repetitive movements; and sleep is disturbed. The patient is unable to perform even light duties and experiences difficulties in daily tasks. This condition may last for months or years.

The early stage is often reversible through work modification and rest breaks. In the later stages, the most important factor is abstinence from performing the causative actions, and rest. This may mean major changes in working habits and life style. Further treatments include physiotherapy, drug administration, and other medical treatments including surgery. Medically, it is important to identify a CTD case early, at a stage that allows effective treatment. Ergonomically, it is even more important to prevent injury by recognizing potentially injurious activities and conditions and alleviating them through work (re-)organization and work (re-)design.

Of course, it is best to avoid conditions that may lead to an overexertion injury. Work object, equipment, and tools used should be suitably designed; instruction on and training in proper postures and work habits are important; managerial interventions such as work diversification (the opposite of job simplification and specialization), relief workers, and rest pauses are often appropriate. Of course, it is most important to "not repeat" possibly injurious motions and force exertions, and to avoid unsuitable postures.

As a general rule, jobs should not require often repeated activities, especially if significant body strength must be exerted. (Of course, for weak body parts — such as the fingers — even a numerically small force of a few Newtons of pinch force may be significant, though this may mean little for the strong shoulder muscles.) The wrist should not be severely flexed, extended, or pivoted sideways but in general should remain aligned with the forearm. The forearm should not be twisted. The elbow angle should be varied. The upper arms usually should hang down along the sides of the body. The head should be kept fairly erect. The trunk should be mostly upright when standing; when sitting for long periods, a tall, well-shaped backrest is desirable to lean against, as least during breaks. While body motions (as opposed to maintained stiff posture) are generally desired, trunk rotation should not be required at work. There should be enough room for the legs and feet to stand or sit comfortably. It is important that the postures of the body segments, and of the whole body, can and will be varied often during the working time: walking is better than standing still.

Avoidance of overexertion, whether by redesign of an existing workstation or by appropriate planning of a new workstation, follows one simple general rule: Let the operator perform "natural" activities for which the human body is suited over prolonged time.

Jobs should be analyzed for their posture, movement, and force requirements, using, for example, the well-established industrial engineering procedure of motion and time study. Each element of the work should be screened for factors that can contribute to CTDs. A convenient tool is the "Flow Chart — Flow Diagram" technique discussed in Key 1. This procedure can be applied on a mini-scale to the minute details of a job as well as for the macro evaluation of a whole facility demonstrated in Key 1. Consider an assembly task, for example: It also consists of operations, transports, inspections, delays, and storages, and probably other special activities (such as reach, twist, lift, sort). The analyst can use this easy and well-understood technique of flow charting/diagramming to determine the good and the undesirable details of a job. (Other applicable analysis techniques are described in ergonomic and industrial engineering textbooks.)[4]

Ergonomic Check 2-3. JOB ANALYSIS

A job analysis can be performed for planned and existing conditions. It should concern especially

- rapid and often repeated body actions
- repetitive exertion of finger, hand, or arm forces
- contorted body joints

so that these can be eliminated by better job design.

A job analysis can show that material handling activities may in fact be quite different from expected. For example, an evaluation of the load-handling activities in the distribution center of a large transport company[5] showed that of 3217 handlings of objects other than boxes,

---

[4] Konz, S. (1995). *Work Design* (4th ed.). Scottsdale, AZ: Publishing Horizons.
[5] Baril-Gingras, G. and Lortie, M. (1995). The Handling of Objects Other Than Boxes: Univariate Analysis of Handling Techniques in a Large Transport Company. *Ergonomics*, 38, 905–925.

- 23% were sliding,
- 22% raising,
- 19% pivoting
- 9% turning
- 8% lowering
- 7% carrying
- 6% voluntary dropping
- 3% rolling
- 2% overcoming a resistance (when objects were caught or jammed).

In 57% of the handlings, the object remained supported in some way on a surface; in 42% it left the surface (lifted, lowered, carried, etc.); in 4% it fell unintendedly.

Each handling was broken into phases: *pre-transfer*, when the object was positioned, often brought closer to the body; followed by *transfer* and *placement*. The final phase was often a *post-placement* when the object was adjusted to its final position. (Note that such descriptive terms can be generated by the analyst as needed for the job analysis.)

Such a breakdown indicates the details of the job, and hence how the task might be redesigned to make it easier and less hazardous. It also provides guidance on how to provide equipment such as roller conveyors that make the material movement least strenuous and fastest (see Key 3). Finally, the analysis also shows which body activities should be avoided or preferred for health and effort reasons, as discussed in Key 5.

After the job analysis has been completed, workstation and equipment can be ergonomically engineered and work procedures organized to avoid unnecessary stress on the operator. Table 2-3 provides an overview of ergonomic measures to fit the job to the person.

The general process of work, the particular hand tools to be used, or the parts on which work needs to be performed, should be designed, or if found wanting, altered as needed to fit human capabilities. The opposite way, i.e., selecting persons who seem to be especially able to perform work that most people cannot do, or to let several people work at the same workstation alternately so that nobody has to work long periods of time on the

Table 2-3 Ergonomic Measures to Avoid Common ODs

CTD	Avoid in general	Avoid in particular	Do	Design
Carpal tunnel syndrome	Rapid, often-repeated finger movements, wrist deviation	Dorsal and palmar flexion, pinch grip, vibrations between 10 and 60 Hz	Use large muscles, but infrequently and for short durations	the work object properly
Cubital tunnel syndrome	Resting forearm on sharp edge or hard surface		—	the job task properly
DeQuervain's syndrome	Combined forceful gripping and hard twisting		Let wrists be in line with the forearm	—
Epicondylitis	"Bad tennis backhand"	Dorsiflexion, pronation		hand tools properly ("bend the tool, not the wrist")
Pronator syndrome	Forearm pronation	Rapid and forceful pronation, strong elbow and wrist flexion	—	
Shoulder tendonitis, Rotator cuff syndrome	Arm elevation	Arm abduction, elbow elevation	Let shoulder and upper arm be relaxed	round corners, pad
Tendonitis	Often repeated movements, particularly with force exertion; hard surface in contact with skin; vibrations	Frequent motions of digits, wrists, forearm, shoulder	Let forearms be horizontal or more declined	place work object properly

Table 2-3 Ergonomic Measures to Avoid Common ODs (Continued)

CTD	Avoid in general	Avoid in particular	Do	Design
Tendosynovitis, DeQuervain's syndrome, Ganglion	Finger flexion, wrist deviation	Ulnar deviation, dorsal and palmar flexion, radial deviation with firm grip	Let wrists be in line with the forearm	the work object properly
Thoracic outlet syndrome	Arm elevation, carrying	Shoulder flexion, arm hyperextension		round corners, pad
Trigger finger or thumb	Digit flexion	Flexion of distal phalanx alone		—
Ulnar artery aneurism	Pounding and pushing with the heel of the hand			the job task properly
Ulnar nerve entrapment	Wrist flexion and extension	Wrist flexion and extension, pressure on hypothenar eminence		hand tools properly
White finger, vibration syndrome	Vibrations, tight grip, cold	Vibrations between 40 and 125 Hz		—
Neck tension syndrome	Static head posture	Prolonged static head/neck posture	Alternate head/neck postures	place work object properly

Adapted from Kroemer, K.H.E. (1992). Avoiding cumulative trauma disorders in shop and office. *J. Inds. Hyg. Assoc.*, 53(9), 596–604. With permission.

same job, are basically inappropriate measures that should be applied only if no other solution can be found.

The 1994 book *Ergonomics*[6] contains several recommendations for job design:

1. Provide a chair with a headrest, so that one can relax neck and shoulder muscles at least temporarily.
2. Provide an armrest, preferably cushioned, so that the weight of the arms need not be carried by muscles crossing the shoulders and elbow joints.
3. Provide flat, possibly cushioned surfaces on which forearms may rest while the fingers work.
4. Provide a wrist rest for people operating traditional keyboards, so that the wrist cannot drop below the key level.
5. Round, curve, and pad all edges that otherwise might be point-pressure sources.
6. Furnish jigs and fixtures to hold work pieces in place, so that the operator does not have to hold the work piece.
7. Place jigs and fixtures so that the operator can easily access the work piece without contorting hand, arm, neck, or back.
8. Select bins and containers, and place them so that the operator can reach into them with the least possible flexion/extension, pivoting, or twisting of hand, arm, and trunk.
9. Furnish tools whose handles distribute pressure evenly over large surfaces of the operator's digits and palm.
10. Select hand tools (and work procedures) that do not require pinching.
11. Select the lightest possible hand tools.
12. Select hand tools that are properly angled so that the wrist need not be bent.
13. Select hand tools that do not require the operator to apply a twisting torque.

[6] Kroemer, K.H.E., Kroemer, H.B., and Kroemer-Elbert, K.E. (1994). *Ergonomics: How to Design for Ease and Efficiency*. Englewood Cliffs, NJ: Prentice Hall.

14. Select hand tools whose handles are so shaped that the operator does not have to apply much grasping force to keep it in place or to press it against a work piece.
15. Avoid tools that have sharp edges, fluted surfaces, or other prominences that press into tissues of the operator's hand.
16. Suspend or otherwise hold tools in place so that the operator does not have to do so for extended periods of time.
17. Select tools that do not transmit vibrations to the operator's hand.
18. If the hand tool must vibrate, have energy absorbing/dampening material between the handle and the hand (yet, the resulting handle diameter should not become too large).
19. Make sure that the operator's hand does not become undercooled, which may be a problem particularly with pneumatic equipment.
20. Select gloves, if appropriate, to be of proper size, texture, thickness.

More than three decades ago, Peres[7] wrote:

It has been fairly well established, by experimental research overseas and our own experience in local industry, that the continuous use of the same body movement and sets of muscles responsible for that movement during the normal working shift (not withstanding the presence of rest breaks), can lead to the onset initially of fatigue, and ultimately of immediate or cumulative muscular strain in the local body area (p. 1)

It is sometimes difficult to see why experienced people, after working satisfactorily for, say, 15 years at a given job, suddenly develop pains and strains. In some cases these are due to degenerative arthritic changes and/or traumatic injury of the bones of the wrist or other joints involved. In other cases, the cause

---

[7] Peres, N.J.V. (1961). Process work without strain. *Australian Factory*, 1, 1–12.

seems to be compression of a nerve in the particular vicinity, as for example, compression of the median nerve in carpal tunnel syndrome. However, it may well be that many more are due to cumulative muscle strain arising from wrong methods of working... (p. 11)

If all opportunities to automate or mechanize material movement have been exhausted, material handling will have to be assigned to people. Job requirements must be established that do not overload the person. One must organize the task, set job procedures, and determine details to enable the human to perform the work safely and efficiently. The main rules are:

- Move loads horizontally.
- Deliver goods at hip height when the person is standing.
- Do all material handling in front of the trunk.
- Make sure the objects are easy to handle.

Ergonomic Check 2-4. MOVE LOADS HORIZONTALLY

If people must move material

- make sure that the material is moved predominantly in a horizontal plane.

Have people push and pull (without twisting and severe bending of the body) rather than lift and lower.

Ergonomic Check 2-5. DELIVER AT HIP HEIGHT

If material is brought to the workplace of a standing person

- make sure that material is delivered at about hip height so that persons don't have to lift it.

Ergonomic Check 2-6.
LIFT AND LOWER IN FRONT OF THE TRUNK

If people must lift or lower material

• make sure that any lifting and lowering is done between hip and shoulder heights.

Lifting and lowering below hip height and above shoulder height, away from the body, and with a sideways twisted body, are most likely to result in an overexertion injury.

Ergonomic Check 2-7. EASILY HANDLED MATERIAL

If people must move material,

• make sure the material is light, compact, and easy to grasp.
• make sure the material does not have sharp edges, corners, or pinch points.

A light object strains the spinal column, muscles, and ligaments less than a heavy object. Compact material can be held closer to the body than a bulky one. A solid object with good handles is more safely held and more easily moved than a pliable one.

Ergonomic Check 2-8. CONTAINER USE

If material is delivered in a container

• make sure the material can be easily removed,

particularly that the operator does not have to "dive" into a container to reach the material, and

• make sure the container does not have sharp edges on which the operator could get injured.

WORK ENVIRONMENT

The work environment contributes to the safety and efficiency of manual material activities if it is ergonomically well designed and maintained.

The *visual environment* should be well lit, clean, and uncluttered, allowing good depth perception, discrimination of visual details, of differences in contrast, and of colors. Kroemer, Kroemer and Kroemer-Elbert (1994)[8] made the following recommendations:

1. Proper vision requires sufficient quantity and quality of illumination.
2. Special requirements on visibility (and particularly the decreased vision abilities of the elderly) require care in the arrangement of proper illumination.
3. Illuminance of an object is inversely proportional to the distance from the light source.
4. Use of colors, if selected properly, can be helpful; but color vision requires sufficient light.
5. What counts most is the luminance of an object, the energy reflected or emitted from it, that meets the eye.
6. Luminance of an object is determined by its incident illuminance and by its reflectance. Reflectance is the ratio of reflected light to received light, in percent.
7. The ability to see an object is much influenced by the luminance contrast between the object and its background, including shadows. Contrast is usually defined as the difference in luminances of adjacent surfaces, divided by the larger luminance, in percent:

$$\text{Contrast (in \%)} = (L_{max} - L_{min})(L_{max})^{-1} \times 100$$

8. Avoid unwanted or excessive glare.
   There are two types of glare: *direct glare*, which meets the eye directly from a light source (such as the headlights of an oncoming car); *indirect glare*, which is light

---

[8] Kroemer, K.H.E., Kroemer, H.B., and Kroemer-Elbert, K.E. (1994). *Ergonomics: How to Design for Ease and Efficiency*. Englewood Cliffs, NJ: Prentice Hall.

reflected from a surface into the eyes (such as headlights of a car in the rearview mirror).

Direct glare can be avoided by:

a. Placing high-intensity light sources outside the cone of 60 degrees around the line of sight.
b. Using several low-intensity light sources instead of one intense source, placed away from the line of sight.
c. Using indirect lighting, where all light is reflected toward a suitable surface (within the luminaire, or at the ceiling or walls of a room) before it reaches the work area. This generates even illuminance without shadows. (But shadows may be desirable to better see objects.)
d. Using shields or hoods over reflecting surfaces, or visors over a person's eyes, to keep out the rays from light sources.

Indirect glare can be reduced by:

e. Diffuse, indirect lighting.
f. Dull, matte, or other unpolished surfaces.
g. Properly distributed light over the work area.

9. Use of *direct lighting* (when rays from the source fall directly on the work area) is most efficient in terms of illuminance gain per unit of electrical power, but it can produce high glare, poor contrast, and deep shadows. The other way is to use *indirect lighting*, where the rays from the light sources are reflected and diffused at suitable surfaces before they reach the work area. This helps to provide an even illumination without shadows or glare.

The *acoustic environment* should

- facilitate speech communications,
- prevent noise-induced hearing loss,
- transmit desired sounds and signals reliably,

- be agreeable and pleasant to the human,
- minimize sound-related annoyance and stress.

The interactions between human hearing capabilities and the sound environment, especially "noise," are well known. Many publications (including Kroemers' (1994) *Ergonomics*[9] and Plog's (1996) *Fundamentals of Industrial Hygiene*[10]) as well as OSHA regulations contain details. In general, the prevailing sound levels should be below 75dBA; warning signals and sounds indicating unusual conditions should be clearly perceptible by the operator. Conditions of high noise can contribute to an overall loading of the operator, and hence affect safety of material handling.

The *thermal environment* can also make load handling safe and efficient, or conversely strenuous and stressful.

The human body generates energy, and at the same time exchanges (gains or loses) energy with the environment. Since a rather constant core temperature must be maintained, suitable heat flow from the body outward must be achieved in a hot climate, and excessive heat loss must be prevented in a cold environment.

The human body has a complex control system to maintain the deep body core temperature very close to 37°C (about 99°F). Changes in core temperature of ±2° from 37°C affect body functions and task performance severely, while deviations of ±6°C are usually lethal. Energy is exchanged with the environment through *radiation, convection, conduction,* and *evaporation.*

Heat exchange through *radiation,* the flow of electromagnetic energy, depends primarily on the temperature difference between two opposing surfaces, for example, between a window pane and a person's skin, and on the exposed skin surface. Heat is always radiated from the warmer to the colder surface. Hence, the body can either lose or gain heat through radiation. This radiative heat exchange does not depend on the temperature of the air between the two opposing surfaces.

---

[9] Kroemer, K.H.E., Kroemer, H.B., and Kroemer-Elbert, K.E. (1994). *Ergonomics: How to Design for Ease and Efficiency.* Englewood Cliffs, NJ: Prentice Hall.
[10] Plog, B.A. (Ed.) (1996). *Fundamentals of Industrial Hygiene* (4th ed.). Itasca, IL: National Safety Council.

Energy is also exchanged through *convection* and *conduction*. In both cases, the heat transferred is again proportional to the areas of human skin participating in the process, and to the temperature difference between skin and the adjacent layer of the external medium. Exchange of heat through convection takes place when the human skin is in contact with air and fluids, e.g., water; and through conduction when in contact with solids. Heat energy is transferred from the skin to a colder medium next to the skin surface, or transferred to the skin if the surrounding medium is warmer. Convective heat exchange is facilitated if the medium moves quickly along the skin surface thus maintaining a temperature differential.

As long as such temperature gradient exists, there is always some natural movement of air or fluid: this is called "free convection." More movement can be produced by forced action (such as by an air fan, or while swimming in water rather than floating motionless): this is called "induced convection."

Heat exchange by *evaporation* is only in one direction: the human loses heat by evaporation; there is no condensation of water on the skin, which would add heat. Evaporation of water requires an energy of about 580 cal/cm^3 of evaporated water, which reduces the heat content of the body by that amount. Some water is evaporated in the respiratory passages but most (as sweat) on the skin. The heat lost by evaporation from the human body depends on the participating wet body surfaces and on air humidity. As with convection, movement of the air layer at the skin increases the actual evaporative heat loss as this replaces humid air by drier air. Some evaporative heat loss occurs in hot as well as cold environments because there is always evaporation of water in the lungs that increases with enlarged ventilation at work, and there is also continuous diffusion of sweat (insensible water loss) onto the skin surface. Heat is produced in the body's metabolically active tissues: primarily at skeletal muscles, but also in internal organs, fat, bone, connective and nerve tissue. Heat energy is circulated throughout the body by the blood. Heat is exchanged with the environment at the body's respiratory surfaces and of course through the skin.

In a *cold* environment, heat must be conserved, which is primarily done by reducing blood flow to the skin and by increasing surface insulation. If the skin and near tissues, especially muscle, become too cold, control of activities, especially of the speed and accuracy of motions, becomes difficult and slow. In a *hot* environment body heat must be dissipated and gain from the environment prevented. This is primarily done by increases in blood flow to the skin, in sweat production, and evaporation. If the body is about to be overheated, internal heat generation must be diminished. Therefore, muscular activities will be reduced, possibly to the extent that no work is being performed anymore. (One is tempted to call this a "physiological siesta.")

The thermal environment at work is determined by four physical factors: air temperature, humidity, air movement, and temperature of surfaces that exchange energy by radiation. The combination of these four factors determines the physical conditions of the climate and of our ability to perform work efficiently and safely. Each of these factors is, in principle, under engineering control, although existing conditions may make ergonomic corrections more or less feasible and economic. Establishing a suitable climate of, say, 18 to 22°C (65 to 70°F) and 50% relative humidity with low air speed is often fairly easy in an enclosed room, for example, by air conditioning; but generating good thermal conditions for material handling outside may be quite difficult. Still, there are means to help: for example, a large fan makes package removal from a truck easier on a hot summer day; infrared radiative heating installed at the same loading dock warms the loader during a cold spell in the winter. Ergonomics handbooks[11] provide much information.

Ergonomic Check 2-9. WORK ENVIRONMENT

Work conditions must fit human vision and hearing capabilities, and allow suitable energy exchange with the environment through a comfortable climate.

Careful *housekeeping* helps to avoid injuries and facilitates efficient work habits. Safe gripping of shoes on the floor, or good body support from the chair when seated, are important conditions for safe load handling. Poor coupling between the shoes

---

[11] For example: Kroemer, K.H.E., Kroemer, H.B., and Kroemer-Elbert, K.E. (1994). *Ergonomics: How to Design for Ease and Efficiency.* Englewood Cliffs, NJ: Prentice Hall; Plog, B.A. (Ed.) (1996). *Fundamentals of Industrial Hygiene* (4th ed.). Itasca, IL: National Safety Council.

and the floor can result in slipping, tripping, or misstepping. Floor surfaces should be kept clean to provide a good coefficient of friction with the shoes. Clutter, loose objects on the floor, dirt, spills, etc. can severely reduce friction and lead to slip-and-fall accidents, and make material handling haphazard, as Figure 2-1 illustrates.

Figure 2-1 Bad housekeeping makes work hazardous. (Courtesy of International Labour Office, 1988.)

Key 3   EQUIPMENT

There are two types of materials that can be moved by equipment: bulk or unit material. Bulk material, stored and moved in large volume, often comes as powders, granules, or liquids. Large containers, vessels, or bulk carriers are used for this material. Of interest here is material that comes in smaller units or as single items. These are handled either singly, or grouped together in bags, packages, cartons, crates, or on pallets. Many different kinds of equipment and machines are available to handle this material.

A reasonable distinction can be made between equipment that provides assistance to the material handler at the workplace and equipment that provides for in-process movement between work stations.

Equipment for assistance at the workplace:

- Lift tables, hoists
- Ball transfer tables, turn tables
- Loading/unloading devices
- Non-powered trucks, walkies, and dollies

Equipment used mostly for in-process movement:

- Powered walkies, rider trucks, tractors
- Conveyors of many kinds, trolleys
- Cranes

Obviously, several of these can be used both at the workplace and for in-process movement, such as hoists, conveyors, and trucks.

Finally, there is a group of material movement equipment primarily used at receiving and in warehousing. This includes:

- Stackers
- Reach trucks
- Lift trucks
- Cranes
- Automated storage and retrieval systems

For automated and highly mechanized machinery that makes it unnecessary for humans to be directly involved, there is a specialized industry that supplies and installs such systems. The Material Handling Institute is one source of related information. In this text we will concentrate on the ability of equipment to relieve persons from strainful load handling, with emphasis on the safety and ease of use of such equipment for the operator.

Often, simple carts and dollies can be used to transport and lift objects. Figures 3-1, 3-2, and 3-3 show equipment that can be employed to bring loads to the workstation and to lift it to the correct working height. Figure 3-3 illustrates the use of electric, hydraulic, and vacuum hoists. Figure 3-4 shows a small crane attached to the bed of a pickup truck; it eliminates the need for the driver to load and unload manually. Figure 3-5 depicts different conveyors on which one can move objects easily from one workstation to the next. Note specifically the flexible roller conveyor, which can be moved from one workstation to another. Figures 3-6, 3-7, and 3-8 show several details of conveyors that help achieve ease of work.

Ergonomic Check 3-1. USE EQUIPMENT NOT PEOPLE

Use equipment instead of persons to move loads

- at the workplace
- in the work process
- for receiving and warehousing.

Figure 3-1 Dollies, carts, and trucks to move and lift loads. (Adapted from Kroemer, K. H. E., Kroemer, H. B., and Kroemer-Elbert, K. E. (1994). *Ergonomics: How to Design for Ease and Efficiency.* Englewood Cliffs, NJ: Prentice Hall. With permission. Also Courtesy of Best Diversified Products, Inc.)

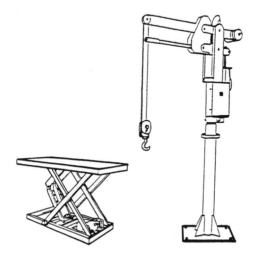

Figure 3-2 Stationary scissors lift table and small crane.

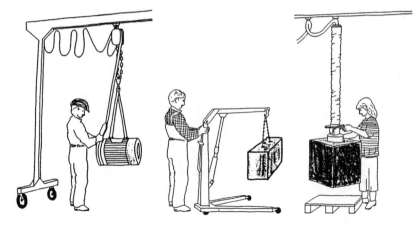

Figure 3-3 Electric, hydraulic, and vacuum hoists. (Adapted from British Standard BS 40004.)

There are many kinds of often inexpensive equipment that can do the holding, turning, carrying, pushing, pulling, lowering, and lifting of loads that would otherwise be performed by persons. However, whether this will indeed be done by machine depends, besides economic considerations, on the layout of the workstation and organization of the work itself. For example: will a lift table be installed next to an assembly workstation if this means removal or relocation of other workstations in order to make sufficient room? Will an operator use a hoist to lift a 20-kg object if this is awkward and consumes more time than to simply grasp it and pull it up — even at the risk of some back pain?

Figure 3-4 Small crane mounted on a truck. (Adapted from British Standard BS 40004.)

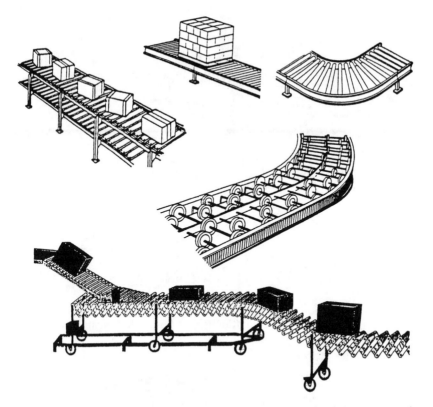

Figure 3-5 Several types of conveyors suitable for moving loads between work-stations. (Adapted from Kroemer, K. H. E., Kroemer, H. B., and Kroemer-Elbert, K. E. (1994). *Ergonomics: How to Design for Ease and Efficiency.* Englewood Cliffs, NJ: Prentice Hall. With permission. Also Courtesy of Best Diversified Products, Inc. and New Dominion Equipment Company.)

Ergonomic Check 3-2. EASY-TO-USE EQUIPMENT

- Is the load handling more easily done with the help of equipment than by hand or does it take longer and appear awkward?

Obviously, facility layout as well as workplace design, must be suitable for the use of equipment. Furthermore, the operator must be convinced (trained) that it is worthwhile to go through the effort of using a hoist instead of heaving the material by

Figure 3-6 Intersection of two conveyors with ball transfer to move the load easily in any direction. (Courtesy of Boston Technical Furniture.)

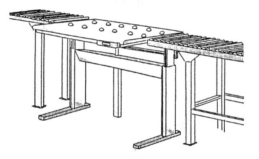

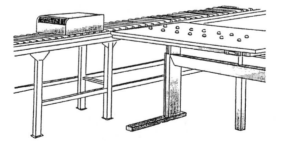

Figure 3-7 Workstation with ball transfers between or adjacent to roller conveyors. (Courtesy of Boston Technical Furniture.)

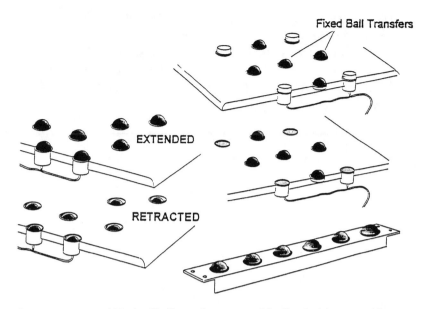

Fixed Ball Transfers

EXTENDED

RETRACTED

Figure 3-8 Several kinds of ball transfers: retractable, fixed with retractable stops, mounted in an add-on strip. (Courtesy of Boston Technical Furniture.)

hand. Considerations related to Key 1, "Facility," Key 2, "Job Design," and Key 4, "People," interact with each other and affect equipment use and selection.

Equipment must not only be selected to be able to perform the material handling job, but must also fit the human operator. It is the ease of use, as well as the safety of the persons working with the equipment, that must be considered. Unfortunately, some material movement equipment such as cranes, hoists, powered and hand trucks, and particularly fork lift trucks have shown an alarming lack of consideration of human factors and safety principles in their design.

> *...I was shocked, dismayed, and perturbed. Recently I attended a regional industrial exhibition that had an emphasis on materials-handling equipment. I intentionally went around looking for bad or lacking human engineering. I found plenty...inadequate labels; wrong-size controls; lack of shape, position, color coding; controls that could be inadvertently actuated; absence of guard rails; unintelligible instructions; slippery surfaces; impossible reach requirements; sharp edges; unguarded pinch-*

points; extreme strength requirements; lack of guards; spaces needed for maintenance too small for the human hand; poorly located emergency switches; and so on, and so on!...Spacecraft and supersonic aircraft and missile monitoring equipment need human engineering; so, too, do hydraulic hoists and forklift trucks and conveyor systems and ladders...

This was written in 1984 by R.B. Sleight on page 7 of the January edition of the *Human Factors Society Bulletin*. Unfortunately, one still finds equipment with many of these deficiencies in use, and some inadequate machinery is still being produced today.

Ergonomic Check 3-3.
ERGONOMIC EQUIPMENT CHECKLIST

Equipment intended to make load handling easier for the operator, or to take over the hard part of the task, should be checked for human engineering by this list.

Does it fit the operator's

- body size and space needs?
- mobility and preferred motions/postures?
- strength and endurance?
- vision and hearing characteristics?
- ability to hear signals?

When selecting equipment, one needs to consider that it must be safe and easy to operate. The old-type forklift truck, depicted in Figure 3-9, was neither: the load obscured the operator's forward view and therefore much of the driving was done rearward, with the driver's body contorted and the controls awkward to operate. The driver's space was cramped and the seat hard, directly transmitting shocks and vibrations to the trunk.

In 1995, usability aspects of simple trucks and trolleys were reviewed in industry and hospitals.[12] From the great variety of

---

[12] Mack, K., Haslegrave, C.M., and Gray, M.I. (1995). Usability of Manual Handling Aids for Transporting Materials. *Applied Ergonomics*, 26, 353–364.

Figure 3-9 Old-time forklift truck: there is a protective cage, but vision, space, and control operation are bad. (From Kroemer, K. H. E., Kroemer, H. B., and Kroemer-Elbert, K. E. (1994). *Ergonomics: How to Design for Ease and Efficiency.* Englewood Cliffs, NJ: Prentice Hall. With permission.)

observations and comments made by the workers, the following primary problem areas were identified:

- *Required operating force*, especially for getting the aid into motion, controlling when in motion, and stopping it.
- *Stability*, mostly with two-wheeled and tall aids (made "tall" either due to device design or by putting on a tall load).
- *Steerability*, mostly related to the number, location, size, and type of wheels; quality of wheel bearings; and the size (wheelbase) of the aid.
- *Handles*, often too low, too small, too short, inappropriately bent and badly located.
- *Starting*, particularly with heavy loads and small wheels on the hand truck, and when the device must be tilted before starting to move.
- *Stopping*, particularly on down-slopes, when cornering, and on wet or slippery floors.

- *Loading and unloading,* especially with pallets that have too low an opening space, making insertion of the fork difficult; and with sack trucks where the front edge of the "shoe" may be difficult to slip under the load.
- *Security of the load,* especially when going over bumps or around corners with aids that by design have no provision to keep the load in place, or with aids equipped with insufficient or awkward-to-use securing devices.

Usability depends not only on the design features of the equipment, but also on

- *Characteristics of the load:* type, size, weight, weight distribution, shape.
- *Use conditions,* such as maintenance (especially of the wheels), frequency and duration of use, loads per trip, space available, perceived "pressure" such as in rush work, availability of assistance, lighting, vibration.
- *Flooring,* type of surface, cleanliness, friction; even, flat, or sloped; bumps, steps.

All of these design aspects and use conditions are in the domain of ergonomics and can and should be controlled to achieve ease and safety of use.

SPECIAL HUMAN ENGINEERING DESIGN RECOMMENDATIONS

Ergonomics and human factors research has provided information on design features that are needed to fit equipment to the human, in particular in order to provide safe and efficient working conditions. Much of this information was originally developed for military applications, but has found its way into industrial settings as well.

Containers and trays must be designed to have the right weight, size, balance, and coupling with the human hand. The container should be as light as possible to add little to the load of the material. Its size and handle arrangement should be so that the center of the loaded tray is close to the body. It should be well balanced, with its weight centered in line with but below the hand-holds.

Ergonomic Check 3-4. FITTING CONTAINERS AND TRAYS

Make sure that containers and trays

- fit the hand in size, weight, balance, and coupling.

On a reasonably sized box, such as 40 × 40 × 40 cm, the best placement for handles is high in the middle of each side, as shown as location "2/2" in Figure 3-10. This is easy to remember and to do. Some other handle locations are also acceptable, especially 3/8 and 6/8.

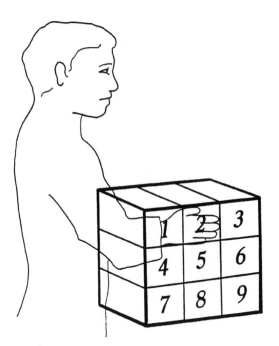

Figure 3-10 Preferred handle locations on a 40 × 40 × 40 cm box. On each side, "location 1" is highest and closest to the handler's body. Handles may be angled so that the wrists are approximately straight when the hand is on the handle of the box. (Adapted from Drury, C. G., Deeb, J. M., Hartman, B., Woolley, S., Drury, C. E., and Gallagher, S. (1989). Symmetric and asymmetric manual materials handling; Part 1: Physiology and psychophysics. *Ergonomics*, 32, 467–489. With permission.)

Coupling with the hands should be facilitated by handholds on both sides. Protruding handles or gripping-notches are best; cutouts, or drawer-pull types, are acceptable. Handholds should be so designed and oriented that hand and forearm of the operator are aligned. Do not force the operator to work with a bent wrist: bend the handle, not the wrist. Carpal tunnel syndrome and other cumulative trauma problems of the wrist are less likely to occur if a person can work with a straight wrist — see also Key 2, "Job Design."

The handle should be of such shape and material that the squeezing forces are distributed over the largest possible hand area. The handle surface in contact with the hand should be rounded, with a diameter in the range of $2^{1}/_{2}$ to 5 cm (1 to 2 inches). Its surface should be slip resistant.

INSTRUCTIONS AND WARNINGS

Written instructions are not necessary if equipment and operation are designed to be perfectly obvious. However, when instructions are needed, the labeling should be done according to these rules:

- *Orientation:* A label and the information printed on it should be oriented horizontally so that it can be read quickly and easily. (Note that this applies if the operator is used to reading horizontally, as in Western countries.)
- *Location:* A label should be placed on or very near the item that it identifies.
- *Standardization:* Placement of all labels should be consistent throughout the equipment and system.
- *Equipment functions:* A label shall primarily describe the function ("what does it do?").
- *Abbreviations:* Common abbreviations may be used. If a new abbreviation is necessary, its meaning should be obvious to the reader. The same abbreviation should be used for all tenses and for the singular and plural forms of a word. Capital letters should be used, periods normally omitted.

- *Brevity:* The label inscription should be as concise as possible without distorting the intended meaning or information. The texts should be unambiguous, with redundancy minimized.
- *Familiarity:* Words should be chosen, if possible, that are familiar to the operator.
- *Visibility and legibility:* The operator should be able to read the label easily and accurately at the anticipated actual reading distances, at the anticipated worst illumination level, and within the anticipated vibration and motion environment. Important are: contrast between the lettering and its background; the height, width, strokewidth, spacing, and style of letters; and the specular reflection of the background, cover, or other components.
- *Font and size:* Typography determines the legibility of written information; it refers to style, font, arrangement, and appearance.

  Font (typeface) should be simple, bold, and vertical, such as Futura, Helvetica, Namel, Tempo, and Vega.

  Recommended height of characters depends on the viewing distance:

  Viewing distance 35 cm, suggested height 22 mm.
  Viewing distance 70 cm, suggested height 50 mm.
  The ratio of strokewidth to character height should be between 1:8 to 1:6 for black letters on white background, and 1:10 to 1:8 for white letters on black background.
  The ratio of character width to character height should be about 3:5.
  For continuous text, mix upper and lower case letters. (For labels, upper case letters only.)

One must keep in mind that many people will read instructions only if their first attempts to operate do not succeed. Also, to be read, instructions must be right there under their eyes (not hidden somewhere on page 384 of a manual) and they must have enough contrast against the background, and luminance even in dim light, to be seen and be readable.

Ergonomic Check 3-5. INSTRUCTIONS

The main rules for instructions on how to use equipment are:

- write in the simplest, most direct manner possible.
- give only the needed information.
- describe clearly the required action. (Never mix different instruction categories, such as warning, operation, and maintenance.)
- use familiar words.
- be brief, but not ambiguous.
- locate labels in a consistent manner.
- words should read horizontally, not vertically.
- label color should contrast with equipment background.

Warnings are really a sign of defeat on the part of the designer who strives to design devices to be perfectly safe. If complete safety cannot be achieved, users must be warned of dangers associated with the use, and one must provide instructions for safe use to prevent injury and damage. It is much preferred to have an active warning, an alerting device that warns the human of an impending danger and tells what to do; but often passive warnings are employed, usually written instructions on labels. Such passive warnings rely completely on the human to recognize the danger and to behave according to the warning and instructions. Table 3-1 lists guidelines for passive warnings. More information on when and how to warn can be found in the literature.[13]

LIFT BELTS

Weight lifters have long been putting various kinds of supportive braces around their midriffs in the hope that the belts

---

[13] Kroemer, K.H.E., Kroemer, H.B., and Kroemer-Elbert, K.E. (1994). *Ergonomics: How to Design for Ease and Efficiency.* Englewood Cliffs, NJ: Prentice Hall; Lehto, M.R. and Miller, J.M. (1986). *Warnings,* Vols. 1 and 2. Ann Arbor, MI: Fuller; Ryan, J. (1991). *Design of Warning Labels and Instructions.* New York: Van Nostrand Reinhold; Laughery, K.R., Wogalter, M.S., and Young, S.L. (1994). *Human Factors Perspectives on Warnings.* Santa Monica, CA: Human Factors and Ergonomics Society.

Ergonomic Check 3-7. WARNINGS

Warnings are a "sign of defeat": the designer was unable to come up with a perfectly safe device. If warnings are needed

- choose active warnings

that alert the human of impending danger and tell what to do. Passive warnings merely rely on the human to realize the risk and to behave accordingly.

Table 3-1 Major Aspects of Warnings

Whom to warn

All potential product users who might be at risk so that they may be educated to protect themselves.

All potential customers (purchasers) whose employees might be at risk so that work practices and workstations can be modified to protect people.

Sales, marketing, and service staff in contact with users and customers.

The general public.

Reasons to warn

One must warn of known or knowable potential danger of injury from normal use of the product.

    The more serious the potential injury, the less obvious the danger, the more insidious the onset of injury, the larger the number of people at risk the greater the duty to warn.

About what to warn

Importance of proper use.

Excessive or improper use, which may cause serious injury of a painful, permanent, or disabling nature.

Individuals who are at particular risk and must take particular care.

The need to take rest breaks.

The need to seek medical attention immediately after any symptoms appear.

How to warn

Sign on the product.

Instructions in Manual.

Software to flash warnings.

Promotional literature and advertising.

Sales and service personnel's instructions to customers and users.

---

allow them to lift heavier loads with less danger of injury, especially to the lower back. Yet, lifting-related overexertions and injuries are still prevalent among weight lifters. In spite of this fact, lift belts of various designs have been promoted for occupational and leisure use.

Enos and Mitchell paraphrased (on page 6 in the August 1993 issue of the Human Factors and Ergonomics Society's Bulletin) stances taken against the use of back braces as follows:

- Having workers stand on their heads before the shift would be as effective, because of the Hawthorne effect.[14]
- Fad, quick-fix, Band-Aid
- Psychological benefits generally outweigh the physical benefits
- "Superman syndrome" — i.e., false sense of security
- Referring to back braces as "ergonomic" is unethical because it is a complete misapplication of the term.
- Back braces do not eliminate the workers' exposure to hazard.
- They treat the symptoms, not the causes.
- The risk of injury is increased off the job, when the belt is not used.
- They promote muscle atrophy.
- Use of back braces is associated with a risk of physical discomfort.
- Companies choose to use them in order to demonstrate, quickly and inexpensively, their wish to prevent back injuries.
- Resources should be better directed toward activities that promote workers' health and morale.

Quite a few controlled experiments, case studies, and testimonials have been published with respect to the following questions: Which (if any) physiological and biomechanical effects do such belts have? How about psychological and attitudinal effects? Does their use decrease the incidence and/or severity of material handling injuries? Can they be used instead of ergonomic design of handling task, procedures, and equipment? Of course, global answers are often difficult to give because of the many and highly diverse conditions under which loads are

---

[14] Hawthorne effect is the positive result of a treatment of people even if the action itself is ineffective; but the people react positively to the show of concern and interest. Also known as the "consultant's best tool."

moved manually. However, several specific studies and surveys of the literature[15] lead to the following general conclusions:

- Lateral twisting and bending motions of the trunk may be reduced, especially at large displacements of loads to the side of the body. However, this benefit could be accompanied by increased trunk muscle tension and hence higher spinal loading.
- In lifting tasks requiring near maximal efforts, belts may increase intra-abdominal pressure which may relieve some spinal compression. However, this benefit may also be accompanied by increased trunk muscle tension, which could augment spinal compression.
- Lift belts seem not to introduce mechanical or motivational advantages to the user.
- The use of supports appears to be marginally effective in reducing the number of injuries, but the intensity and cost of treating people injured while wearing a belt appears to be higher than for those injured without belt.
- Using lift belts of any kind does not afford more protection than proper material handling procedures without belt.

Ergonomic Check 3-9. LIFT BELTS ARE NOT ENOUGH

- Back belts are not a substitute for proper ergonomics of material handling.

---

[15] Lavender, S.A. and Kenyeri, R. (1995). Lifting belts: a psychophysical analysis. *Ergonomics,* 38, 1723–1727; Lavender, S.A., Thomas, J.S., Chang, D., and Andersson, G.B.J. (1995). Effects of lifting belts, foot movement, and lift asymmetry on trunk motions. *Human Factors.* 37, 844–853; McGill, S.M. (1993). Abdominal belts in industry: a position paper on their assets, liabilities and use. *Am. Ind. Hyg. Assoc. J.,* 54, 752–754; Mitchell, L.V., Lawler, F.H., Bowen, D., Mote, W., Asundi, P., and Purswell, J. (1994). Effectiveness and cost-effectiveness of employer-issued back belts in areas of high risk for back injury. *JOM Journal of Occupational Medicine,* 36, 90–94.

People come in all sizes: weak and strong, fit and untrained, female and male, young and old. Ergonomic knowledge of human characteristics stems from several disciplines, such as:

- Anthropometry: body dimensions, motion capabilities, muscularity
- Work physiology: capabilities for physical work
- Biomechanics: mechanical structure and strain behavior of the body
- Industrial psychology: attitudes and behaviors
- Sociology: people perception within society

The first three topics will be discussed in the following text in a survey fashion. More detail is available for the designer, engineer, and manager in books on applied (engineering, work, exercise) physiology.[16]

BODY SIZES — ANTHROPOMETRY

Adults come in many different body sizes: short and tall, light and hefty.[17] Body sizes change with age, nutrition, and training.

---

[16] For example: Astrand, P.O. and Rodahl, K. (1986). *Textbook of Work Physiology* (3rd Ed.) New York: McGraw-Hill; Eastman Kodak Company (Ed.) (Vol. 1, 1983, Vol. 2, 1986); *Ergonomic Design for People at Work.* New York: Van Nostrand Reinhold; Kroemer, K.H.E., Kroemer, H.J., and Kroemer-Elbert, K.E. (1997). *Engineering Physiology: Bases of Human Factors/Ergonomics* (3rd ed.). New York: Van Nostrand Reinhold; Kroemer, K.H.E., Kroemer, H.B., and Kroemer-Elbert, K.E. (1994). *Ergonomics: How to Design for Ease and Efficiency.* Englewood Cliffs, NJ: Prentice Hall; Mellerowicz, H. and Smodlaka, V.N. (1981). *Ergonometry.* Baltimore, MD: Urban and Schwarzenberg.

[17] Only in Garrison Keillor's Lake Wobegon are "all women strong, all men good-looking, and all children above average."

Throughout history, attempts have been made to establish simple rules that describe body dimensions. Among these is the idea that one could define an average person who might be used as a standard design template. This approach was and is wrong: there are no persons who are average in all or even many respects. People of average height may not have an average weight, have no average arm length, are not of average strength, and do not want to be treated as average.

Ergonomic Check 4-1.
THE "AVERAGE PERSON" PHANTOM

The "average person" appears only in statistics.

• Do not design for a ghost called "Average."

Therefore, equipment, workstations, and material to be handled must be designed and controlled to *suit and fit the whole range of body dimensions, and capabilities, of the people who handle material.* Handle sizes must fit small and large hands. Working surfaces must be arranged for short as well as tall people. Even relatively weak people must be able to move material, while the strongest should not break it.

Ergonomic Check 4-2. DESIGN FOR RANGES

Equipment, workplaces, and materials must be carefully designed and adapted

• to fit all users' body sizes and physical capabilities

from small to big, from strong to weak.

Table 4-1 presents body dimensions describing the U.S. adult population. Only 5% of all people have dimensions that are smaller than the 5th percentile value given. Conversely, only 5% have body dimensions larger than the 95th table values. Obviously, below the 50th percentile are 50% of all data and above it, the other 50%.

Table 4-1 Body Dimensions of U.S. Civilian Adults, Female/Male, in cm

	Percentiles			Std. deviation
	fem. 5th male	fem. 50th male	fem. 95th male	fem. S male
Heights				
Stature ("Height")f	152.78/164.69	162.94/175.58	173.73/186.65	6.36/6.68
Eye heightf	141.52/152.82	151.61/163.39	162.13/174.29	6.25/6.57
Shoulder (acromial) heightf	124.09/134.16	133.36/144.25	143.20/154.56	5.79/6.20
Elbow heightf	92.63/99.52	99.79/107.25	107.40/115.28	4.48/4.81
Wrist heightf	72.79/77.79	79.03/84.65	85.51/91.52	3.86/4.15
Crotch heightf	70.02/76.44	77.14/83.72	84.58/91.64	4.41/4.62
Height (sitting)s	79.53/85.45	85.20/91.39	91.02/97.19	3.49/3.56
Eye height (sitting)s	68.46/73.50	73.87/79.20	79.43/84.80	3.32/3.42
Shoulder (acromial) Ht (sitting)s	50.91/54.85	55.55/59.78	60.36/64.63	2.86/2.96
Elbow height (sitting)s	17.57/18.41	22.05/23.06	26.44/27.37	2.68/2.72
Thigh height (sitting)s	14.04/14.86	15.89/16.82	18.02/18.99	1.21/1.26
Knee height (sitting)f	47.40/51.44	51.54/55.88	56.02/60.57	2.63/2.79
Popliteal height (sitting)f	35.13/39.46	38.94/43.41	42.94/47.63	2.37/2.49
Depths				
Forward (thumbtip) reach	67.67/73.92	73.46/80.08	79.67/86.70	3.64/3.92
Buttock-knee distance (sitting)	54.21/56.90	58.89/61.64	63.98/66.74	2.96/2.99
Buttock-popliteal distance (sitting)	44.00/45.81	48.17/50.04	52.77/54.55	2.66/2.66
Elbow-fingertip distance	40.62/44.79	44.29/48.40	48.25/52.42	2.34/2.33
Chest depth	20.86/20.96	23.94/24.32	27.78/28.04	2.11/2.15
Breadths				
Forearm-forearm breadth	41.47/47.74	46.85/54.61	52.84/62.06	3.47/4.36
Hip breadth (sitting)	34.25/32.87	38.45/36.68	43.22/41.16	2.72/2.52

Table 4-1 Body Dimensions of U.S. Civilian Adults, Female/Male, in cm (Continued)

	Percentiles			Std. deviation
	fem. 5th male	fem. 50th male	fem. 95th male	fem. S male
Head dimensions				
Head circumference	52.25/54.27	54.62/56.77	57.05/59.35	1.46/1.54
Head breadth	13.66/14.31	14.44/15.17	15.27/16.08	0.49/0.54
Interpupillary breadth	5.66/5.88	6.23/6.47	6.85/7.10	0.36/0.37
Foot dimensions				
Foot length	22.44/24.88	24.44/26.97	26.46/29.20	1.22/1.31
Foot breadth	8.16/9.23	8.97/10.06	9.78/10.95	0.49/0.53
Lateral malleolus height[f]	5.23/5.84	6.06/6.71	6.97/7.64	0.53/0.55
Hand dimensions				
Circumference, metacarpal	17.25/19.85	18.62/21.38	20.03/23.03	0.85/0.97
Hand length	16.50/17.87	18.05/19.38	19.69/21.06	0.97/0.98
Hand breadth, metacarpal	7.34/8.36	7.94/9.04	8.56/9.76	0.38/0.42
Thumb breadth, interphalangeal	1.86/2.19	2.07/2.41	2.29/2.65	0.13/0.14
Weight (in kg)	39.2*/57.7*	62.01/78.49	84.8*/99.3*	13.8*/12.6*

* Estimated by Kroemer.

Note: In this table, the entries in the 50th percentile column are actually "mean" (average) values. The 5th and 95th percentile values are from measured data, not calculated (except for weight). Thus, the values given may be slightly different from those obtained by subtracting 1.65 S from the mean (50th percentile), or by adding 1.65 S to it.

[f] above floor; [s] above seat.

Adapted from U.S. Army data reported by Gordon, C.C., Churchill, T., Clauser, C.E., Bradtmiller, B., McConville, J.T., Tebbetts, I., and Walker, R.A. (1989). *1988 Anthropometric Survey of U.S. Army Personnel: Summary Statistics Interim Report.* (Natick/TR 89-027). Natick, MA: United States Army Natick Research, Development and Engineering Center.

Statistically, in a normal distribution the 50th percentile corresponds to the mean "m" value, also called the average. (In statistics texts this is often designated by the symbol x̄.) The standard deviation "S" is a measure of the spread of the data. A larger standard deviation indicates that the data are more spread out to the extremes, while a small standard deviation indicates that all data are bunched together closely at the mean.

Figure 4-1 shows a typical result of measurements of one anthropometric variable, for example, stature (standing height). The recorded data are distributed symmetrically on each side of the 50th percentile value. As one divides this data distribution into sections of one standard deviation each, known percentages are included in these sections. Hence, one can describe a normal distribution of data using very simple statistics: about 68% of all measured values lie within ± 1 standard deviation from the mean, while 95% of all data points are located within the boundaries established by 2 standard deviations below and above the mean.

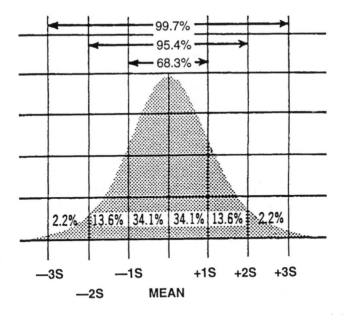

Figure 4-1 Anthropometric data are usually distributed in a "normal (Gauss) curve." Because of this distribution, it is possible to state the percentage that lies between +1 and –1 standard deviations S, or between any two other points determined from multiples of S units.

Table 4-2   "Factor k" Values for the
            Computation of Percentiles
            of a Normally Distributed
            Population

For percentiles		Use factor k
1st	or 99th	2.33
2nd	or 98th	2.06
3rd	or 97th	1.88
5th	or 95th	1.65
10th	or 90th	1.28
16.5th	or 83.5th	1.00
20th	or 80th	0.84
25th	or 75th	0.67
50th	or 50th	0

In fact, any percentage cutoff point can be calculated by using the mean value and a multiple of the standard deviation. Multiplication factors are presented in Table 4-2. Percentiles between 0 and 50 are calculated from $m - k \cdot S$; percentiles between 50 and 100 are calculated from $m + k \cdot S$.

Percentiles can be used to describe clearly and easily any distribution of anthropometric data. For example: Table 4-1 shows that 5% of all American women are shorter than 152.8 cm, 95% are taller (k = 1.65, see Table 4-2). Their mean (average) height is 162.9 cm; 50% are shorter, and 50% taller (k = 0). Two out of ten females are taller than 168.3 cm (k = 0.84). Selecting the appropriate values for k, one can establish cutoff points that relate to percentiles anywhere along the distribution of body dimensions.

Table 4-2 lists the multiplication factors with which we can calculate percentile values: One needs only the mean (same as average, same as 50th percentile value), the standard deviation S, and the multiplication factors k. In the examples just used, the factors k are 1.65, 0.00, and 0.84, respectively; S is 6.36 cm and m is 162.94 cm.

Ergonomic Check 4-3. IT'S EASY TO GET PERCENTILES

Using a simple statistical procedure, one can calculate any percentile value p:

- $p = m + k \cdot S$

Using this approach, and the data displayed in Tables 4-1 and 4-2, dimensions can be looked up or calculated for the design of tools, equipment, and furniture, etc., that one wants to fit to given users. Here are some examples:

What is the height of a working surface for standing male and female operators? Assuming that this height coincides with "Elbow Height, Standing" (see Table 4-1), one finds that the mean elbow height for females is 99.8 cm, and for males 107.3 cm. Since "design for the average" is not acceptable, one has to decide for what range to design.

- *Example A.* Assume one wants to have low benches to fit small females, i.e., the 5th percentile. Table 4-1 shows that in this case "elbow height" is 92.6 cm.
- *Example B.* Assume high benches are to be designed to fit fairly tall males. If the cutoff point is the 84th percentile, then the elbow height to be considered is 112.1 cm (i.e., 107.3 + 1.0 × 4.8).
- *Example C.* Given the above values, one would choose to use a height range of 92.6 to 112.1 cm, an adjustment range of nearly 20 cm. The height of all benches should be increased by the height of the shoe heels worn, say 2 cm. But people are likely to "slump" while working: this will probably offset that 2-cm increase for heels. The height of the objects at which the work is performed above the table surface should be estimated in order to adjust (lower) the bench height accordingly.

Ergonomic Check 4-4. MAX, MIN, or RANGE

Which design strategy applies?

- ONE design cutoff: either a minimum or a maximum value; or
- A RANGE extending between a lower and an upper percentile

for fitting workstations, equipment, and work tasks to the people involved in material handling.

For some jobs the viewing distance is important, if eye control of the operation is necessary. In this case, "eye height, standing" would be an additional design variable, to be considered together with the desired viewing distance and viewing angle.

THE BODY AS ENERGY SOURCE — WORK PHYSIOLOGY

The human body must maintain an energy balance between external demands, caused by the work and the work environment, and the capacity of internal body functions to produce that energy. The body is an "energy factory," converting chemical energy derived from nutrients (protein, carbohydrate, fat, and alcohol) into externally useful energy. Final stages of this process take place at skeletal muscles.

The energy conversion needs oxygen transported from the lungs by the blood. Also, the blood removes by-products generated in the energy conversion, such as carbon dioxide, water, and heat, which are dissipated in the lungs, where, of course, oxygen is absorbed into the blood. Heat is also expelled through the skin, much of it by sweat. The blood circulation is powered by the heart.

Thus, the pulmonary system (lungs), the circulatory system (heart and blood vessels), and the metabolic system (energy conversion) establish central limitations of a person's ability to perform strenuous work.

A person's capability for labor is limited also by muscular strength, by the ability for movement in body joints (e.g., the knees) or by the spinal column (e.g., "weak back"). These are local limitations for the force or work that a person can exert.

Very often local limitations establish the upper limits for performance capability. For example, one may simply lack the strength to lift an object because the hands are too far extended in front of the body. The "mechanical advantages" at which muscles must work often determine one's ability to perform a given job.

While handling material, the force exerted with the hands must be transmitted through the whole body, that is via wrists, elbows, shoulder, trunk, hips, knees, ankles, and feet to the floor. In this chain of force vectors, the weakest link determines the capability of the whole person to do the job. If muscles are weak, or if they have to work at mechanical disadvantages, the handling force is reduced. Often the weakest link in this chain of forces is the low back region, particularly at the lumbar spine. Muscular or ligament strain and painful displacements of the vertebrae and/or of the intervertebral discs in the spine may limit a person's ability for material handling.

Models have been developed to describe the central and local limitations just discussed. Simplified for convenience, they may be grouped as follows:[18]

Physiological models provide information on energy conversion and expenditure (measured in kilocalories) and on the loading of the circulatory system (e.g., heartbeats per minute).

Muscle strength models provide information on the ability to exert force on an outside object. Up to now, these measurements were usually done with the muscle kept at a given length (isometric), i.e., with involved body segments in frozen positions. Isokinematic (isokinetic) and isoinertial techniques are applied to assess dynamic muscle strength, that is, while the body moves.

Biomechanical models provide information on strain tolerance and performance capabilities of the body, particularly of the spinal column and its discs, and on the effect of body positions on available muscle strength.

Psychophysical models combine physical measures with subjective assessments of the perceived strain. They provide synergistic judgments of material handling capabilities and limitations. Psychophysical models are not discussed in this text.

The Physiological Model

In many respects one may compare the way in which the body generates energy with the functioning of a combustion engine: fuel (food) is combusted for which oxygen must be present. The combustion yields energy that moves parts (of the

[18] Kroemer, K.H.E., Kroemer, H.J., and Kroemer-Elbert, K.E. (1997). *Engineering Physiology: Bases of Human Factors/Ergonomics* (3rd ed.), New York: Van Nostrand Reinhold.

engine or body) mechanically. The fueling and cooling system (blood vessels) moves supplies (oxygen, carbohydrates, and fat derivatives) to the combustion sites (muscle, other organs) and removes combustion by-products (carbon dioxide, water, heat) for dissipation (at skin and lung surfaces).

The processes are controlled by the complex and overlapping central nervous system, the hormonal system and the limbic system. Control centers (in the brain and spinal cord) rely on feedback from various body parts and provide feedforward signals according to general (autonomic, innate, learned) principles and according to situation-dependent (voluntary, motivational) rules.

The *respiratory system* provides oxygen for the energy metabolism and dissipates metabolic by-products. The *circulatory system* carries oxygen from the lungs to the consuming cells to which it also brings the fuel, i.e., derivatives of carbohydrates and fats. Furthermore, it removes metabolic by-products from the combustion sites. The *metabolic system* provides the chemical processes in the body, particularly those that yield energy. These three systems must work closely together, and a person's ability to perform physically demanding work depends on this cooperation and of course on the proper and efficient functioning of each.

The respiratory system moves air to and from the lungs, where part of the oxygen contained in the inhaled air is absorbed into the blood stream; it also removes carbon dioxide, water, and heat from the blood into the air to be exhaled.

The volume of air exchanged in the lungs depends on the "heaviness" of the work performed. At rest, one breathes 10 to 20 times every minute. In light exercise, primarily the "tidal volume" of each breath is increased, but with heavier work the respiratory frequency also quickly increases up to about 45 breaths per minute. This indicates that breathing frequency, which can be measured easily, is not a reliable indicator of the heaviness of work performed.

The respiratory system is able to increase its moved air volume and absorbed oxygen by large multiples. The minute volume can be increased from about 5 L/min to 100 L/min or more; that is an increase in air volume by a factor of 20 or more. Though not exactly linearly related to it, the oxygen consumption shows a similar increase.

The *circulatory system* carries oxygen from the lungs to the cells, where nutritional materials, also brought by circulation from the digestive tract, are metabolized. Metabolic by-products (CO_2, heat, and water) are dissipated by circulation.

The circulatory system is nominally divided into two subsystems: the systemic and the pulmonary circuits, each powered by one half of the heart (which can be considered a double pump). The left side of the heart supplies the systemic section, which branches from the arteries through the arterioles and capillaries to the metabolizing organ (e.g., muscle); from there it combines again from venules to veins to the heart's right side. The pulmonary system starts at the right ventricle, which powers the blood flow through pulmonary artery, lungs, and pulmonary vein to the left side of the heart.

The heart is in essence a hollow muscle that produces, via contraction and with the aid of valves, the desired blood flow. The ventricle is filled through the valve-controlled opening from the atrium. The heart muscle contracts (*systole*) and when the aortic valve opens, much of the blood is ejected from the ventricle into the systemic system. Then, the aortic valve closes with the beginning of the relaxation (*diastole*) of the heart, while the elastic properties of the aortic walls propel the stored blood into the arterial tree, where elastic blood vessels smooth out the waves of blood volume. At rest, about half the volume in the ventricle is ejected (stroke volume) while the other half remains in the heart (residual volume). With increasing work load, the heart ejects a larger portion of the contained volume and increases its contraction frequency. At a heart rate of 75 beats per minute, the diastole takes less than 0.5 seconds and the systole just over 0.3 seconds; at a heart rate of 150 beats per minute, the periods are close to 0.2 seconds each. Hence, an increase in heart rate occurs mainly by shortening the duration of the diastole.

The heart has its own intrinsic control system, which operates, without external influences, at resting rates that differ from person to person: usually between 50 to 70 beats per minute. Increased heart rate is controlled by the central nervous system in response to greater demand for blood flow to muscles (and other organs) by physical exertion. A heart rate of 120 beats per minute sustained over a prolonged time indicates a rather heavy work load even for a healthy person.

Since the available blood volume of about 5 liters in an adult does not vary appreciably, that volume must be pumped very quickly through the circulatory system, especially when the work demands are high. This means increased demands on the heart as the pump in the system. Cardiac output can be affected by two factors: the frequency of contraction (heart rate) and the pressure generated by each contraction in the blood. Both determine the so-called cardiac *minute volume*. The cardiac output of an adult at rest is around 5 L/min. When performing strenuous exercise, this level might be raised five times to about 25 L/min, while a well-trained athlete may reach up to 35 L/min.

A healthy heart can pump much more blood through the body than usually needed. Hence, a circulatory limitation is more likely to lie in the transporting capability of the vascular portions of the circulatory system than in the heart itself. Blood must seep through the consuming organ (e.g., a muscle) via very fine blood vessels. The pressure differential from the arterial side (incoming) to the venous side (outgoing) maintains the movement of blood through the "capillary bed." Here, both flow velocity and blood pressure are low, allowing nutrients and oxygen to enter the extracellular space of the tissue, at the same time permitting the blood to accept metabolic by-products from the tissue. Compression of the capillary bed can occur if the striated muscle itself contracts strongly, at more than about 20% of its maximal capability. If such contraction is maintained, the muscle hinders or shuts off its own blood supply and cannot continue the contraction. Thus, sustained strong static contraction is self-limiting, as we all experience when working overhead with our arms extended upward. *One should not require anybody to work in sustained contractions, be it in keeping the body in position or in grasping a handle tightly. Instead, equipment and job should be so designed that frequent changes in muscle tension are facilitated, best by body movement.* This is the reason why ergonomists exhort to avoid static posture but encourage dynamic motion.

Heart rate generally follows oxygen consumption and hence energy production of the dynamically working muscle in a linear fashion from moderate to rather heavy work. However, static (isometric) muscle contraction leads to a higher heart rate, apparently because the body tries to bring blood to the tensed muscles. Also, working in a hot environment causes a higher heart rate than at a moderate temperature — see Key 2.

The *metabolic system* is balanced in a healthy body: over days, energy input equals output. The input is from nutrients, from which chemically stored energy is liberated during the metabolic processes within the body. The output is mostly heat and work. Work is measured in terms of physically useful energy, i.e., energy transmitted to outside objects. The amount of such external work performed strains individuals differently, depending on their physique and training. This is again an indication of the close interaction between the metabolic, circulatory, and respiratory systems as sketched in Figure 4-2.

The balance between energy input I (via nutrients) and outputs can be expressed by a simple equation:

$$I = H + W + S$$

where H is the generated heat, W the work performed on an outside object, and S the energy storage in the body (negative if lost from it).

The measuring units for energy (work, heat) are Joules (J) or calories (cal) with 4.2 J = 1 cal. (Exactly: 1 J = 1 Nm = 0.2389 cal = 10^7 ergs = 0.948×10^{-3} BTU = 0.7376 ft lb.) One uses the kilocalorie, kcal or Cal = 1000 cal, to measure the energy content of foodstuffs. The units for power are 1 W = 1 Js^{-1} or 1 kcal hr^{-1} = 1.163 W.

Assuming for simplicity no change in energy storage and also that no heat is gained from or lost to the environment, one can simplify the energy balance equation to

$$I = H + W$$

Human energy efficiency e (work efficiency) is defined as the ratio between work performed and energy input:

$$e = (W/I)100 = [W/(H + W)]100, \text{ in percent}$$

In everyday activities, only about 5% or less of the energy input is converted into work, which is energy usefully transmitted to outside objects. By far the largest portion of the energy input is used for maintaining the body and is finally converted into heat that the body must dissipate into the environment — often a major problem in a hot climate, as mentioned in Key 2.

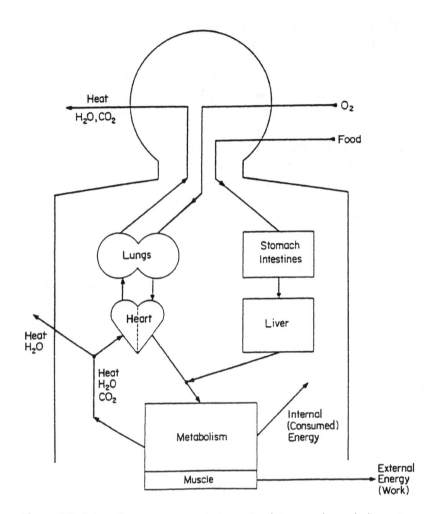

Figure 4-2 Interactions among respiratory, circulatory, and metabolic systems. (From Kroemer, K.H.E., Kroemer, H.J., and Kroemer-Elbert, K.E. (1997). *Engineering Physiology: Bases of Human Factors/Ergonomics,* 3rd edition. New York: Van Nostrand Reinhold. With permission.)

Work (in the physical sense) is done by skeletal muscles. They move body segments against external resistances. The muscle is able to convert chemical energy into physical work or energy. From resting, a healthy human can increase the energy generation up to fifty-fold. Such an enormous variation in metabolic rate not only requires quickly adapting supplies of nutrients and oxygen to the muscle but also generates large amounts of waste products (mostly heat, carbon dioxide, and water) that must be removed. Thus, while performing physical work, the ability to maintain the internal equilibrium of the body is largely dependent on the circulatory and respiratory functions that serve the involved muscles. Among these functions, the control of body temperature is of particular importance. This function interacts with the external environment, particularly with temperature and humidity, as discussed in Key 2.

Energy is supplied to the body as food or drink. Our food consists of various mixtures of organic compounds (foodstuffs) and water, salts, minerals, vitamins, etc., and of fibrous material (mostly cellulose). This roughage does not release energy, which is derived from carbohydrates, fats, and proteins. Their average nutritionally usable energy contents per gram are: 4.2 kcal for carbohydrate, 9.5 kcal for fat, and 4.5 kcal for protein. Alcohol yields about 7 kcal per gram.

Under normal nutritional and exertion conditions, glucose and glycogen are the primary and first-used sources of energy at the cell level, particularly in the central nervous system and at muscles, whereas fat accounts for most of the stored energy reserves. This means an energy storage in the form of fat of about 85,000 kcal for a thin, lightweight person, and nearly 250,000 kcal for a heavy person. In contrast, we have only a few thousand kilocalories available from glycogen and glucose.

The ability to perform physical work is different from person to person and depends on gender, age, body size, fitness, health, environment, and motivation. Commonly, it is measured by that person's oxygen consumption during a standardized test, often on a bicycle ergometer. Assuming an average caloric value of oxygen of 5 kcal per consumed liter of oxygen, one can calculate the energy conversion capability of the person from the volume of oxygen actually consumed.

Heart rate (as a primary indicator of circulatory functions) and oxygen consumption (representing the metabolic conversion taking place in the body) have a linear and reliable relationship in the range between light and heavy work. Therefore, one often can simply substitute heart rate measurement for oxygen measurement. This is a very attractive shortcut since heart rate measurement can be performed easily.

Use of heart rate has a major advantage over oxygen consumption as indicator of metabolic processes: it responds faster to work demands, hence indicates more easily quick changes in body functions due to changes in work requirements. More information on techniques to assess metabolic processes with commercially available equipment can be found in several widely available books.[19]

Ergonomic Check 4-6.
A CLASSIFICATION OF THE "HEAVINESS" OF WORK

Light work	2.5 Cal/min	90 beats per minute
Medium	5 Cal/min	100 beats per minute
Heavy	7.5 Cal/min	120 beats per minute
Very heavy	10 Cal/min	140 beats per minute
		1 Cal = 1 kcal = 1000 cal

The Biomechanical Model

One can understand and explain some characteristics of the human body in mechanical terms: "biomechanics" has been applied to the statics and dynamics of the human body, to explain effects of vibrations and impacts, to explore characteristics of the spinal column, and to the use of prosthetic devices, to mention just a few examples.

Treating the human body as a mechanical system entails gross simplifications, such as disregarding mental functions. Still, many components of the body can be considered in terms of analogies, such as:

Bones — lever arms, structural members
Articulations — joints and bearing surfaces
Tendons — cables transmitting muscle forces

---

[19] See, for example: Astrand, P.O. and Rodahl, K. (1986). *Textbook of Work Physiology* (3rd ed.). New York: McGraw-Hill; Eastman Kodak Company (Ed.) (Vol. 1, 1983; Vol. 2, 1986), *Ergonomic Design for People at Work*. New York: Van Nostrand Reinhold; Kroemer, K.H.E., Kroemer, H.B., and Kroemer-Elbert, K.E. (1994). *Ergonomics: How to Design for Ease and Efficiency*. Englewood Cliffs, NJ: Prentice Hall; Kroemer, K.H.E., Kroemer, H.J., and Kroemer-Elbert, K.E. (1997). *Engineering Physiology: Bases of Human Factors/Ergonomics* (3rd ed.), New York: Van Nostrand Reinhold; Mellerowicz, H. and Smodlaka, V.N. (1981). *Ergometry*. Baltimore, MD: Urban and Schwarzenberg.

Tendon sheaths — pulleys and sliding surfaces
Muscles — motors, dampers, or locks
Organs — generators or consumers of energy

The human skeleton is composed of 206 bones held together in their associated articulations by connective tissues: muscles, tendons, ligaments, and cartilage. One important function of human skeletal bone is to provide an internal framework for the whole body. The long, more or less cylindrical bones that connect body joints are of particular interest to the biomechanic. They are the lever arms at which muscles pull.

Some bony joints have no mobility left, such as the seams in the skull of an adult; some have very limited mobility, such as the connections of the ribs to the sternum. Joints with "one degree of freedom" are simple hinge joints, like the elbow or the distal joints of the fingers. Other joints have two degrees of freedom, such as the wrist joint. Some joints have three degrees of freedom, such as shoulder and hip joints.

The spinal column is a complex structure. It consists of 24 movable vertebrae held together in cartilaginous joints of two different kinds: in front the fibrocartilage disks between the main bodies of the vertebrae, and at the rear the synovial facet joints, two to each vertebra. The spine transfers compressive and shear forces, and both bending and twisting moments from the head and shoulder bones to the pelvis. It also protects the spinal cord, which runs through openings at the posterior (spinal canal) carrying signals between the brain and all sections of the body. This complex rod, transversing the trunk and keeping the shoulders separated from the pelvis, is held in delicate balance by ligaments connecting the vertebrae and by muscles that pull along the posterior and lateral sides of the spinal column. Longitudinal muscles located along the sides and the front of the trunk also both balance and load the spine.

The spine is capable of withstanding considerable loading yet flexible enough to allow a large range of postures. There is, however, a trade-off between load carried and flexibility. If there is no external load on the spine, only its anatomical structures (joints, ligaments, and muscles) restrict its mobility. Applying load to the spinal column reduces its mobility until, under heavy load, the range of possible postures is very limited.

A major load on the spine is compression that results from the pulling force of trunk muscles, which counteract the weight of the upper body and, of course, the external load when handling material. Yet, owing mostly to the slanted arrangement of load-bearing surfaces at discs and facet joints, the spine is also subjected to shear. Furthermore, the spine must withstand both bending and twisting torques.

The spinal column is often the location of discomfort, pain, and injury because it transmits so many internal and external strains. For example, when standing or sitting, impacts and vibrations from the lower body are transmitted primarily through the spinal column into the upper body. Conversely, forces and impacts experienced through the upper body, particularly when working with the hands, are transmitted downward through the spinal column to the floor or seat structures that support the body.

Low back pain is the result of disorders that have been with humans since ancient times: It has been diagnosed among Egyptians 5000 years ago, and was discussed in 1713 by Ramazzini. Everyone has an 8-in-10 chance of suffering from back pain some time during one's life, especially when handling material. A theory popular in recent decades is that many overexertion injuries of the spinal column can be traced to, or explained by, compression of the spinal discs. Such excessive compression is thought to damage, temporarily or permanently, the fibrocartilage disc, which, in consequence, may lead to the intrusion of disc structures into its surroundings, particularly toward the spinal cord. Yet, experimental measurements of the compression within the disc in the living human body are difficult to perform.

Thus, many assessments rely on calculations using biomechanical models that, in turn, often simply assume static compression strains.[20]

Skeletal muscles are the engines that move and stabilize the skeletal structures of the body. They do so by pulling on the bones to which they are attached, thereby creating torques or moments around the body joints, which serve as pivots. Muscles perform

[20] Chaffin, D.B. and Andersson, G.B.J. (1991). *Occupational Biomechanics* (2nd ed.). New York: Wiley.

their functions by contracting, that is, by quickly and reversibly developing internal lengthwise tension, often but not always accompanied by a shortening and followed by elongation.

Under a no-load condition, in which no external force applies and no internal contraction occurs, the muscle is at its resting length. Stimulation with no external load causes the muscle to contract to its smallest possible length, which is at about 0.6 times resting length. In this position, the muscle cannot develop any active contraction force. Stretching the muscle beyond resting length by application of an external force lengthens it passively, which it resists like a rubber band. At about 1.6 times resting length, the muscle shows maximal stretch resistance; further elongation breaks it, or its tendon, or the attachment to bone. Thus, the (isometric) tension developed within a muscle is the sum of active and passive events: the tension is near zero at approximately 0.6 times resting length, about 100% at resting length, and then increasing steeply to about 1.6 times resting length. This explains why we stretch (preload) muscles for a strong force exertion, as in bringing the arm behind the shoulder before throwing.

Newton's First Law states that a mass remains at uniform motion (which includes being at rest) until acted upon by unbalanced external forces. The Second Law, derived from the first, indicates that force is proportional to the acceleration of a mass. The Third Law states that action is opposed by reaction.

The correct unit for force measurement is the newton; one pound-force unit is approximately 4.45 newtons, and 1 kg-force (formerly occasionally called 1 kilopond, kp) equals 9.81 N. The pound (lb), ounce (oz), and gram (g) are usually not force but mass units.

The "stick person" concept consisting of links and joints, embellished with volumes and masses and driven by muscles, can be used to model human motion and strength capabilities. Figure 4-3 shows a model of a (massless) human exerting forces

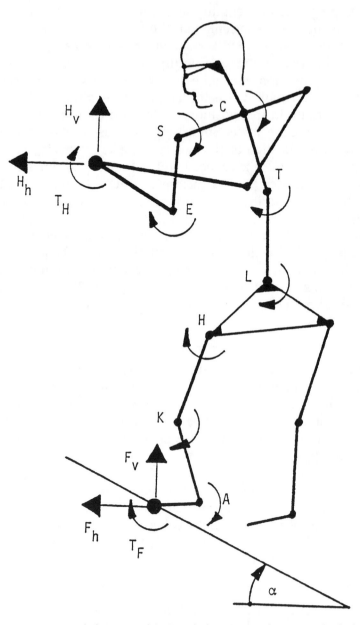

Figure 4-3　Human skeleton simplified as "links connected in joints" (H for hand, E for elbow, S for shoulder, etc.).

or torques with the hand to an outside object. Within the body, the forces are transmitted along the links. First, the force exerted with the right hand, modified by the existing mechanical advantages, must be transmitted across the right elbow (E). Similarly, the shoulder (S) must transmit the same effort, again modified by existing mechanical conditions. In this manner, all subsequent joints transmit the effort exerted with the hands throughout the trunk, hips, and legs and finally from the foot to the floor. Here, force and torque vectors can be separated again into their component directions, similar to the vector analysis at the hands. Still assuming a massless body, the same sum of vectors must exist at the feet as was found at the hands. (Of course, the assumption of no body mass is unrealistic. This can be remedied by incorporating information about mass properties of the human body.)

To consider body motions, instead of the static position assumed so far, complicates the model. Dynamic muscular efforts are much more difficult to describe and control than static contractions. (Since the static case is so easy to control and measure, much of the classic information on muscle strength applies to this "no motion" effort where muscle length remains the same: namely, isometric.) In dynamic activities, muscle length changes, and therefore involved body segments move. Thus, displacement is present, and its time derivatives (velocity, acceleration, and jerk) must be considered. This is a much more complex task for the experimenter than that encountered in static testing. Only recently has a systematic breakdown into independent and dependent experimental variables been presented for dynamic and static efforts. In particular, accelerations (changes in velocity) generate additional forces (e.g., tangential, centrifugal, Coriolis) and torques that may be substantially larger than those experienced in static conditions. Several publications provide further information.[21]

[21] Chaffin, D.B. and Andersson, G.B.J. (1991). *Occupational Biomechanics* (2nd ed.). New York: Wiley; Kroemer, K.H.E., Kroemer, H.J., and Kroemer-Elbert, K.E. (1997). *Engineering Physiology: Bases of Human Factors/Ergonomics* (3rd ed.). New York: Van Nostrand Reinhold; Kroemer, K.H.E., Marras, W.S., McGlothlin, J.D., McIntyre, D.R., and Nordin, M. (1990). Assessing human dynamic muscle strength. *Int. J. Ind. Erg.*, 6, 199–210; Oezkaya, N. and Nordin, N. (1991). *Fundamentals of Biomechanics.* New York: Van Nostrand Reinhold.

Ergonomic Check 4-7. BOTTLENECKS

Different capabilities can limit a person's capability to perform strenuous material handling:

- local limitations are often in muscles and joints.
- central limitations are often in circulation and metabolism.

Load handling may impose demands on one or several of these.

Key 5 TRAINING

Numerous attempts have been made to train material handlers to do their work (particularly lifting) in a safe manner.[22] Unfortunately, hopes for significant and lasting reductions of overexertion injuries after training in safe lifting have been generally disappointed. There are several reasons for the disappointing outcomes of training attempts[23]:

- If the job requirements are stressful, "doctoring the symptoms" (such as by behavior modification) will not eliminate the inherent risk. Designing a safe job is basically better than training people to behave safely.
- People tend to revert to previous habits and customs if practices trained to replace previous ones are not reinforced and refreshed periodically.
- Emergency situations, the unusual case, the sudden quick movement, an increased body weight, or reduced physical well-being may overly strain the body, since training did not include these conditions.

Thus, unfortunately, training for safe material handling (not limited to the popular "lifting") should not be expected to really solve the problem — although, if properly applied and periodically reinforced, training should help to alleviate some aspects.

[22] Garg, A. (1993). What basis exists for training workers in "correct" lifting techniques? In W.S. Marras, W. Karwowski, J.L. Smith, and L. Pacholski (Eds.), *The Ergonomics of Manual Work* (pp. 89–91). London: Taylor and Francis.

[23] Kroemer, K.H.E., McGlothlin, J.D., and Bobick, T.G. (Eds.) (1989). *Manual Material Handling: Understanding and Preventing Back Trauma.* Akron, OH: American Industrial Hygiene Association; Kroemer, K.H.E. (1992). Personnel training for safer material handling. *Ergonomics,* 35, 1119–1134.

Management and health care personnel have every reason, ethical and monetary, to avoid mishandling of objects and causes for injury to people and, naturally, no load handler wants to be injured either. Yet, even well-intended training programs are ineffective unless the trainee becomes personally convinced that safe material handling techniques are available, feasible, and that it is truly in one's own interest to learn and follow them.

Ergonomic Check 5-1. TRAINING MOTIVATION

The key factors for motivating compliance in efforts to prevent injuries while handling loads are

- recognizing the problem and the employee's personal role in it,
- gaining knowledge about one's own capabilities and limitations,
- understanding the relevance of the problem to one's own activities at work, leisure, and home,
- being able to alter the behavior by learning "proper" techniques,
- consistently using these new practices.

WHOM TO TRAIN

Clearly, the material handler is the person who should be trained to handle objects and his or her own body prudently to avoid overexertion, injury, and accidents. But the supervisor, foreman, manager, and engineer direct, assign, and design the tasks and procedures, objects, and work places; therefore, they need similar training so that they understand what may or may not be expected in handling loads. Furthermore, they are the persons who mostly set the stage for attitudes and behavior in the shop or construction site in terms of safe "efficiency" or reckless "get the job done." Making employees aware of authority concern is an underlying theme. There is enough anecdotal evidence to support the common sense opinion that training the boss is about as important as training the worker.

WHO IS THE TRAINER?

The next question is: who should be the trainer? The classic apprentice approach can be quite successful if the experienced worker is safety conscious, follows suitable practices, and is able to teach this behavior to the new worker. However, if this is not the case, then the newcomer is likely to pick up false cues.

Somebody from the outside is often chosen as trainer because that person may have special knowledge and teaching experience, or so that a newcomer is not stuck with bad local customs, or because the outsider is respected as a convincing expert. A major disadvantage can be that the consultant may not be sufficiently familiar with the local conditions and hence could be dismissed as not relevant.

WHAT TO TEACH

A basic question is: what to teach? Certainly, the content of a training course is dependent on the aims of the training. The purpose of training can be specialized or general. Special topics may be related to one or just a few activities, for example, lifting bulky boxes or handling patients in a hospital. General-purpose training often includes the need for "taking care of oneself" by recognizing the importance of the personal role in the material handling activities in choosing proper practices and in staying fit and healthy.

To improve *physical fitness* is one of the general approaches of training workers to prevent injury, especially low back injury. Material handling is a physical job, and it is reasonable to assume that many aspects of physical fitness, such as musculoskeletal strength, aerobic capacity, or flexibility may be associated with the ability to perform load handling tasks without injury. Exercise has been advocated for many years, although its exact role and effectiveness are not completely understood.

Musculoskeletal strength is one of the aspects of physical fitness that is generally believed to be related to back injury. One would intuitively expect that weaker workers have more musculoskeletal injuries than stronger persons, but there is the unexpected opposite finding that "strong" persons may be more often injured than their weaker co-workers.[24] One explanation for the counterintuitive finding is that there exist many different strengths and weaknesses (static and dynamic, of connective tissues, of local and central processes — as discussed in Key 4) that have differing relations to the various demands on the body while handling loads. Although the concept of strength training within an industrial environment as a means of injury prevention is of great interest, no major research literature on this topic seems to exist.

Flexibility, particularly of the trunk, is apparently needed for the bending and lifting activities that are part of load handling. Therefore one might consider stretching exercises useful as a means of warm-up before work and as a starting point for fitness improvement, yet Nordin[25] found flexibility measures to be poor predictors of back problems.

Regaining and improving fitness including flexibility while recovering from a back injury or other disability has always been of concern to patients and to their nurses and physical therapists. This back school concept has been incorporated in "work hardening" where specific body abilities deemed necessary to perform the job are improved through purposefully designed exercises. Fitness training for prevention in back injury is viewed with great interest; however, there still is insufficient evidence on which to make sound judgments on the effectiveness of this approach in general, or on specific programs.

VARIOUS TRAINING METHODS

Although the aim of injury prevention is the same in each case, the methods of how to achieve that aim are quite different. The traditional approach of training in specific lifting techniques alone does not appear effective, probably because there is no one technique that is appropriate for all lifts. The method of preventing injury by increasing knowledge of the body and promoting attitude changes so that workers feel responsible for their bodies

[24] Battie, M.C., Bigos, S.J., Fisher, L., Hansson, T.H., Jones, M.E., and Wortley, M.D. (1989). Isometric lifting strength as a predictor of industrial pain reports. *Spine*, 14, 851–856.
[25] Nordin, M. (1991). Worker training and conditioning, in *Proceedings, Occupational Ergonomics: Work Related Upper Limb and Back Disorders.* San Diego, CA: American Industrial Hygiene Association, San Diego Section.

Ergonomic Check 5-2. TRAINING CONTENT

Determine the purpose of training. Major topics are:

- training of specific lifting techniques, i.e., skill improvement,
- teaching biomechanics, awareness of the body's capabilities and self-responsibility for injuries, thereby changing attitudes, and
- training the body via physical fitness so that it is less susceptible to overexertion.

These aims can be combined.

---

appears quite appealing. However, exactly what should be taught and how is still open. What method is most effective? How much knowledge is needed?

In an experiment with twelve novice lifters, Barker and Atha[26] found that simply providing written guidelines, such as those commonly used in industry, resulted in poorer performance than allowing people to lift in their untrained state.

EVALUATION OF TRAINING SUCCESS

How does one judge the effectiveness of training? The most commonly used methods rely on "objective" data derived from company records on cost and productivity, and from medical records to compare quantitative measures before and after training. The use of company loss data is most common, but sometimes the exact meaning of these numbers is not clear, particularly if other actions take place during the period of data recording that may have had effects on the loss statistics. "Subjective" data result from asking trainees or managers, "Was the training worthwhile?" The value of such judgments varies from case to case and may not carry over to other circumstances.

It would not make sense to do an evaluation of the long-term effectiveness of training if the students never learned the mate-

---

[26] Barker, K.L. and Atha, J. (1994). Reducing the biomechanical stress of lifting by training. *Applied Ergonomics*, 25:6, 373–378.

rial in the first place. How to measure whether the workers have actually absorbed what was presented in training sessions? Using some test or demonstration as a criterion measure of learning directly following the training could be used as an evaluative tool, but what measure to use and how well must one do on the measure to be considered trained (or having a changed attitude) has not been discussed in the literature.

How about the retention of learned information by the trainees after training? There may have been sufficient original learning, but why then is there usually an increase in injuries (or whatever measure used) as time passes? Is this a reflection of good training that is forgotten over time, or was the training not good to start with? Refresher courses probably should be offered, yet the appropriate time intervals between training sessions must be established.

TRAINING FORMAT

Once it has been determined what should be taught, attention must be paid to the form of the course itself. Many courses are taught in a lecture format, with practice of the techniques under discussion at some time during the session. Films and videotapes, audiotapes, posters, and cards in paycheck envelopes have also been used. Programmed instruction, computer-aided instruction, or interactive video are at hand. The (relative) effectiveness of these methods has not been determined.

GENERAL OR CUSTOM-TAILORED TRAINING

Where should the sessions be held? Would a classroom with desks and chairs be more appropriate than a lecture hall? The worksite appears most appropriate, but it may not be suitable for instructional purposes. Is it best to train employees working together as a group, or should the group be split up? Of course, in any situation, there are practical constraints that might limit the ability to implement training ideals, but current information is not available on which to base sound judgment.

Low back disabilities (due to pain and/or injury) are a major health problem among industrial workers. They are the most frequent and most expensive musculoskeletal disorder in the

United States, accounting for at least 25% of all compensable overexertion injuries. Back pain has been related to weak trunk musculature, muscular fatigue, degenerative diseases, and to improper posture and inappropriate lifting techniques; however, the majority of industrial back injuries has not been associated with objective pathological findings and almost every second back pain episode cannot be linked to a specific incident.[27]

One of the problems in training for safe load handling is that broad generalizations concerning material handling tasks across all work environments may not be possible. Task characteristics and requirements differ much among industries (e.g., tire-making, mining, nursing) as well as within one industry, depending on the specific job, handling aids and equipment available, successful implementation of worker selection, and ergonomic job design. Even the group characteristics of material handlers in different industries might be important in designing a training program; for example, hospital workers might have higher educational skills than heavy-industry workers. Female employees are predominant in certain industries or occupations; this can influence training because, as a group, women are about two-thirds as strong as men.

It appears that training should be custom-tailored to the exact conditions in the shop, hospital, construction site, or moving company of concern. Objects, tasks, activities, and material handlers are probably quite different in each, and accordingly different training goals and techniques must be employed. For example: an analysis of the handling of objects other than boxes in a large distribution center[28] showed that

- *asymmetry* of activities was frequent in (type and direction of) upper limb and trunk motions,
- *body efforts* did follow rather consistent patterns but were rarely in a single plane,
- *body flexion* was frequent in the trunk but seldom in the knees,

[27] Kroemer, K.H.E. (1992). Personnel training for safer material handling. *Ergonomics*, 35, 1119–1134; Leamon, T.B. (1994). Research to reality: a critical review of the validity of various criteria for the prevention of occupationally induced low back pain disability. *Ergonomics*, 37, 1959–1974.

[28] Baril-Gingras, G. and Lortie, M. (1995). The handling of objects other than boxes: univariate analysis of handling techniques in a large transport company. *Ergonomics*, 38, 905–925.

- *handlings consisted of several phases and body motions*, not just one motion,
- *achieving a secure coupling* between hand and object was a major determinant of choosing the type of handling procedure.

These findings indicate not only that transfer of the same training techniques from one facility to the next may be inappropriate, but also that a job analysis (see Key 2) might be advisable because it can indicate that changes in work details and equipment (see Key 3) are warranted before the training can commence.

WHO SHOULD PARTICIPATE IN TRAINING?

In the United States, "only" about two of every one hundred employees report a back injury per year. This poses another problem regarding the effectiveness and cost of back care instructions. Of the actually reported injuries, about every tenth is serious, yet these few serious injuries cause by far the largest portion of the total cost. Hence, if one wanted to prevent specifically these serious injuries, two of every one thousand employees would be the target sample, while all thousand must participate in the educational program. Even if one wanted to address all persons who may suffer from *any* kind of back problem, about twenty out of one thousand, this is still a rather expensive approach, which may not appear cost-effective to the administrator.

GUIDELINES

Given the scarcity of information, hardly any training guidelines are well supported by controlled research. This leaves much room for speculation, guesswork, and charlatanry regarding the best way to train people for the prevention of injuries related to material handling. Yet, a few time-proven guidelines are at hand, especially for lifting objects:

The "leg lift" has been heavily promoted and used in training efforts, as opposed to the "back lift." It is indeed, normally, better to straighten the bent legs while lifting rather than unbending the back. But leg lifts can be done only with certain loads, gen-

---

Ergonomic Check 5-3. TRAINING DESIGN

The major factors in training for safe load handling are:

- training whom: load handlers; also supervisors, managers, engineers...
- trainers: experienced workers, in-house safety and health specialists, outside experts...
- individual or group training...
- voluntary or enforced participation...
- training special skills: lifting, lowering, carrying...
- training general abilities: awareness, fitness, biomechanics...
- training settings: at the workplace, in the shop, in the classroom...
- training types: lectures, demonstrations, videos, written materials, posters...
- when to train: before first activity, after some time on the job; repetitions...
- assessment of training outcomes: by incidences, costs, savings, cost/benefits, by subjective assessments...
- measurement of training success; how often during training, how often after being back on the job.

---

erally small ones that fit between the legs. Large and bulky loads can usually not be lifted by unbending the knees; if one attempts to do so, one may in fact stress the lower back more. Hence, proper task and material design are necessary to permit leg lifts.

There are no comprehensive and sure-shot rules for safe lifting. Lifting and other manual material handling are very complex combinations of moving body segments, changing joint angles, tightening muscles, and loading the spinal column. Here are some guidelines for proper lifting:

- *Design manual lifting (and lowering) out of the task and workplace.* If it needs to be done by a person, it should be performed between hip and shoulder height.
- *Be in good physical shape.* If not accustomed to lifting and vigorous exercise, do not attempt to do difficult lifting or lowering tasks.

- *Think before acting.* Place material conveniently. Have handling aids available. Make sure sufficient space is cleared.
- *Get a good grip on the load.* Test the weight before trying to move it. If it is too bulky or heavy, get a mechanical lifting aid or somebody else to help, or both.
- *Get the load close to the body.* Place the feet close to the load. Stand in a stable position, have the feet pointed in the direction of movement.
- *Involve primarily straightening of the legs in lifting.*

Here are some things to avoid:

- Do NOT twist the back or bend sideways.
- Do NOT lift or lower awkwardly.
- Do NOT hesitate to get help, either mechanical or from another person.
- Do NOT lift or lower with arms extended.
- Do NOT continue heaving when the load is too heavy.

Figures 5-1 through 5-4 illustrate some of these rules. It is important to first "pull in" the load before attempting to lift and then to keep the load close to the body. Instead of lifting a load onto the shoulders it is better to lift and carry with a yoke. Of course, use of a conveyor or some other technical instead of manual solution would be better — see Keys 3 and 7.

For more information on different approaches to training, see, for example,

Ayoub, M. M. and Mital, A. (1989). *Manual Materials Handling* — especially Chapter 8. London: Taylor & Francis.
Chaffin, D. B. and Andersson, G. B. J. (1991). *Occupational Biomechanics* (2nd ed.) — especially Chapter 13. New York: Wiley.
Ergonomics, Special Issues Nos. 7, 8, and 9, of Volume, 35, July to September 1992.
Kroemer, K. H. E., Kroemer, H. B., and Kroemer-Elbert, K. E. (1994). *Ergonomics: How to Design for Ease and Efficiency.* — especially Chapter 10. Englewood Cliffs, NJ: Prentice-Hall.

Figure 5-1 Don't attempt to lift a load in front of the knees but try to start the lifting between the feet. (Courtesy International Labour Office.)

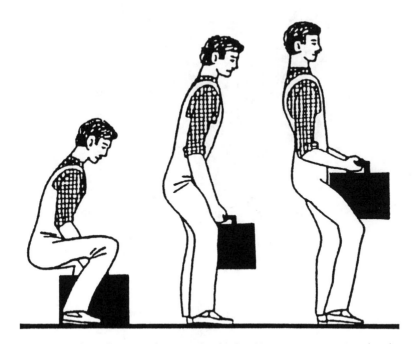

Figure 5-2 Lift and carry close to the body. (Courtesy International Labour Office.)

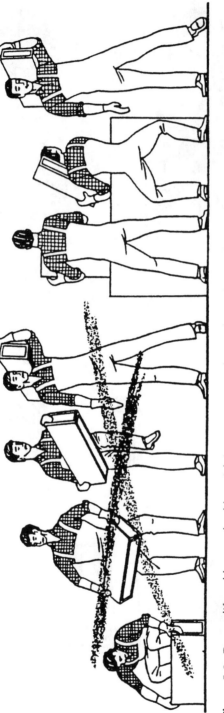

Figure 5-3   Do not lift and lower a load but pick it up, carry it, and set it down close to the trunk. (Courtesy International Labour Office.)

Figure 5-4 Lifting and carrying on a yoke is better than on the shoulder. (Courtesy International Labour Office.)

Key **6**   SCREENING MATERIAL HANDLERS

Screening of material handlers, that is, the selection of persons suitable for stressful material handling, is an important and difficult task. It is important because its success determines whether or not a person can expect to perform load handling safely and in good health, or whether this individual is at risk of an overexertion injury. It is difficult because techniques and procedures are still being developed in research and experimentation, and because care must be taken to avoid discriminatory implications. (Selection and training become much less critical if the work requirements can be reduced through proper job design.)

Ergonomic Check 6-1. MATCHING

Matching job demands with a person's capability requires that both

- individual capabilities or limitations and
- the related job demands

are quantitatively known.

In the past, selection procedures and techniques have been applied that were primarily based on one or several of the following approaches[29]:

---

[29] Kroemer, K.H.E., McGlothlin, J.D., and Bobick, T.G. (Eds.) (1989). *Manual Material Handling: Understanding and Preventing Back Trauma.* Akron, OH: American Industrial Hygiene Association.

The *epidemiological approach* analyzes the circumstances and characteristics that were present in previously observed incidents. For example, certain industries or professions or specific activities have been found to be associated with frequency and severity of overexertion injuries. Overall, while the epidemiological approach provides general information regarding accident probabilities, it does not yield information suitable for individual selection. It is, for example, not feasible to associate such factors as gender or age with certain industry work requirements in order to select individuals for their material handling capability.

The *medical examination* tries to identify an individual's impairments that would make this person vulnerable to overexertion injury risks or, conversely, to identify persons who are healthy and able to perform material handling. While obvious abnormalities (for example, of the spinal column or of the musculature) are likely to be identified in a medical examination, more subtle impairments often remain undetected. Furthermore, the medical examination is usually not specific to the job requirements. Hence, general medical exams as selection procedures can (beyond the obviously unfit) not identify persons who are unable to perform material handling because of increased risk factors, nor can they identify persons who can perform material handling safely.

The *physiological approach* (often part of the physical exam) tries to identify physiological capabilities, or limitations, regarding load handling. As discussed, central limitations are unlikely to be the limiting factors in today's material handling, while local limitations are often not specific enough to be detected in the physiological exam.

The *muscle strength-measuring approach* attempts to assess a person's muscular strength capability with respect to the work demands. While this is a logically appealing approach, its shortfalls are mainly in that specificity, which does not take other capability aspects (such as mobility) into account. Furthermore, previous attempts to measure muscle strength were mostly limited to the measurement of isometric (static) muscle contraction, which is not a sufficiently complete and realistic assess-

ment of the muscular abilities actually required in manual material handling. Measurements of dynamic muscle strength are becoming feasible now and may alleviate
that shortcoming.

The *biomechanical approach* considers the human body simply as a system that can be understood and measured
in mechanical terms. Mathematical and computerized
modeling of the mechanical properties of the body is
often used. Inputs into current models are mostly body
position details, muscular strength information, and
assumptions about spinal column compression capabilities. The validity of the results depends not only on the
quality of the underlying biomechanical model, but also,
obviously, on the inputs to the model. If these inputs are
limited (for example, to isometric muscle strength and
to compression force of the spinal disc), the results of
the modeling approach also will be of limited value, that
is for these specific conditions only, while not applicable
to dynamic work, for example.

The *psychophysical method* is based on the concept that
human capabilities are synergistically determined by
bodily, perceptual, and judgmental capabilities. The psychophysical approach relies on a person's ability to "rate
the perceived strain" stemming from the physical work
performed and to judge whether or not the strain is
personally acceptable for safe performance of that task.
Psychophysics have been successfully applied in many
areas, e.g., in developing scales of effective temperature,
of loudness (noise), of brightness, and of vibration
acceptability. Recent developments have resulted in
scales of acceptable effort for military applications and
for manual material handling, including carrying, pushing, pulling, lowering, and lifting.

The 1991 NIOSH Lifting Guide[30] contains a discussion of
these approaches and combines their results in guidelines for the
design of lifting work. While these guidelines are a much needed
step in the right direction, and are vastly superior to previous

---

[30] Waters, T.R., Putz-Anderson, V., Garg, A., and Fine, L.J. (1993). Revised
NIOSH equation for the design and evaluation of manual lifting tasks. *Ergonomics*, 36(7), 749–776.

simplistic statements of single safe weights for men and women, one needs to consider that ongoing research can quickly and substantially improve the guidelines or in fact replace them with more appropriate and reliable recommendations.

PERSONNEL SELECTION TECHNIQUES

It is reasonable and legitimate to measure a person's ability to perform physically straining work, such as load handling, in order to assure that this individual will not be overexerted. Such selection or screening techniques have been employed during recent years and are being developed further in several research laboratories.

The underlying assumption for all these techniques is that the crucial job requirements are known and that human capabilities related to these requirements can be measured. In comparing job demands with individual capabilities, one expects to find that persons are safe from overexertion if their capabilities exceed demands. The closer one's abilities come to the job demands, the smaller the safety margin. If a person's capability barely matches the job requirements, an overexertion risk exists. This risk is obvious when capabilities fall below demands. (Note, however, that the job demands are often very complex, requiring many dynamic skills that go far beyond simple isometric muscle strength.)

Various screening techniques have been developed in the past, relying on the different approaches just discussed. Three techniques stand out among them:

Isometric strength testing measures the person's ability to exert isometric (static) muscle strength in several frozen body postures. This is a well-established testing procedure relying on biomechanical (static) modeling. It is mainly applicable to force exertions where the body does not move or only moves very slowly and smoothly.

Isokinematic testing measures lift strength capabilities while body segments move at constant (linear or angular) speed.[31]

Isoinertial testing establishes the maximal amount of mass (weight) that a person can lift and lower. The testee grasps

[31] Kroemer, K.H.E., Marras, W.S., McGlothlin, J.D., McIntyre, D.R., and Nordin, M. (1990). Assessing human dynamic muscle strength. *Int. J. Ind. Erg.*, 6, 199–210.

handles, located about 5 cm above the floor, lifts them overhead, then lowers them again. The handles are part of a carriage that is guided between vertical rails. Weights are attached to the carriage so that the testee cannot see them; they are increased until the maximal weight is reached that the person can lift to her or his overhead reach height.[32] This technique has been used widely by the U.S. Army and Air Force and should be applicable to dynamic manual material handling in industry.

There are several grave and largely unresolved problems with the screening of applicants for material handling jobs. The foremost one is: Why is the task so demanding? Couldn't the requirement be lowered so that anybody of normal health and strength can do the job? In fact, maintaining unreasonable task demands and subsequent selection of those who can do the task may be deemed discriminatory — and it certainly is not ergonomic. The second set of major problems concerns the identification of the critical job demands and their quantification. Quite often a job is hard and current occupants try to find another one — but what exactly makes it so hard? If we do not know the exact job demands, we cannot test applicants appropriately. This generates the third set of problems, namely that the examining physician or nurse too often must resort to testing only generic health aspects and performance capabilities of a job applicant while job-specific evaluations would be more meaningful and protective.

Screening procedures are meant to place the right worker in the right job. For this, they must fulfill several requirements: of course, they must be *safe*, which is not always easy to guarantee because the testing is meant to detect the upper limit of a person's performance capability. Tests must be *objective* as opposed to subjective: the outcome may not depend on the whims of the tester. They must be *repeatable*, bringing about the same results when done again. Criteria of *validity* were explained by Chaffin and Andersson[33] in 1984: tests must be *accurate* in that each test

[32] Ayoub, M.M., Jiang, B.C., Smith, J.L., Selan, J.L., and McDaniel, J.W. (1987). Establishing a physical criterion for assigning personnel to U.S. Air Force jobs. *Am. Ind. Hyg. Assoc. J.*, 45, 464–470; Kroemer, K.H.E. (1983). An isoinertial technique to assess individual lifting capability. *Human Factors*, 25, 493–506.

[33] Chaffin, D.B. and Andersson, G.B.J. (1984). *Occupational Biomechanics* (especially pages 399, 400). New York: Wiley.

yields a true measure of what it is designed to assess. This includes that they must be *sensitive* in identifying those persons who indeed will develop a disease or suffer an injury from among all persons who have that risk factor. This is often expressed in terms of the numbers of true positives (those who actually do suffer from disease or injury) compared to false negatives (those who also suffer, but were not predicted to do so). The tests also must be *specific*, which indicates their accuracy in identifying those who indeed will not develop a problem; these are known as true negatives.

$$\textit{Sensitivity (in \%)} = 100 \cdot \# \text{ (true positives)} / \# \text{ (true positives + false negatives)}$$

$$\textit{Specificity (in \%)} = 100 \cdot \# \text{ (true negatives)} / \# \text{ (true negatives + false positives)}$$

Finally, the tests must have *predictive* capability in being able to indicate with good certainty the true positives from all who test positive. It depends on both sensitivity and specificity and on the prevalence of the disease or injury. This is often called *positive predictive value*.

$$\textit{Predictive Value (in \%)} = 100 \cdot \# \text{ (true positives)} / \# \text{ (true positives + false positives)}$$

Ergonomic Check 6-2. SCREENING AIMS

To be able to place the right person on the right job, the screening procedure must be

- safe
- objective
- repeatable
- accurate.

ERGONOMIC DESIGN OF WORKPLACE AND WORK TASK

We all handle material daily. We push, pull, lift, lower, hold, and carry while moving, packing, and storing objects, while loading and unloading machinery. The objects may be soft or solid, bulky or small, smooth or with corners and edges. We may handle one object or perhaps several at a time, and they may be by themselves or come in bags, boxes, or containers, with or without handles. We may handle material occasionally or repeatedly, often as part of our jobs, but also during leisure.

Handling material can involve risks of injury or health hazards, particularly if the objects are too heavy, too bulky, cannot be grasped securely, or if they must be handled too often, or require awkward postures and body motions.

Ergonomic Check 7-1. OBJECTS TO BE HANDLED

Ergonomic design of objects to be handled, of the containers in which they come, and of the workstations helps to avoid overexertions and injuries. The major object rules are:

- have compact objects and packages that can be manipulated securely,
- which are of light weight,
- and which can handled in front of the trunk.

Manipulation of even light and small objects can strain us because we have to stretch, move, bend, or straighten body parts, using fingers, arms, trunk, and legs. Heavy loads pose additional strain on the body owing to their weight or bulk, or lack of handles. Exerting force on an object with the hands strains wrists, elbows, shoulders, the trunk, and particularly the lower back and the legs. The directions and magnitudes of the internal forces are different, depending on body posture, the location of the object, and the direction of force that must be exerted on it. The primary area of physiological and biomechanical concern has been the lower back, particularly the discs of the lumbar spine.

Ergonomic Check 7-2. BODY PARTS AT RISK

All body parts may be overly strained in material handling, but the

- low back
- shoulders
- arms, and
- hands

are particularly at risk.

Compression of the vertebral discs is well researched and guidelines have been developed for proper material handling based primarily on that knowledge, as discussed. Yet, other strains on the spinal column include shear, twist, and bending, as well as strains on muscles and ligaments. For these body structures, less definite information is available, but ongoing research should provide ergonomic information in the near future.

The loading of the musculoskeletal system of the body may come from activities done to an external object, such as in pulling or pushing, lifting or lowering, carrying or holding. Thus, the body strain may be static or dynamic, of fast onset, and of short or long duration. The material handling may occur just once during a day or often. The strains depend on body posture, and where in relation to the body force must be applied to the object, and in what direction.

Ergonomic Check 7-3. STRAIN ON BODY FUNCTIONS

All energy and force needed to handle an object must be developed within the body.

- Energy development strains the load handler's metabolic and circulatory capabilities.
- If very high force must be exerted, a person's musculoskeletal capabilities set the limits.

If the body generates force in motion, the acceleration determines the amount of musculoskeletal force needed, according to Newton's Second Law: force = mass × acceleration. Thus, simply setting a weight limit (such as in kg or lb) does not do enough to prevent overexertion injuries.

The location of the object to be handled with respect to the body (see Keys 1, 2, and 3) is a major determinant of the needed musculoskeletal effort, as are the magnitude and direction of required force.

Ergonomic Check 7-4. LOCATING THE OBJECT

- The object should be located directly in front of the trunk of the body.
- Heavier objects should be at about hip height; lighter objects may be at belly or chest height.
- Forces should be required either in fore or aft direction, or up or down.

The loads must be close and within easy reach of the hands in front of the trunk. Bending and stooping, particularly sideways twisting of the body while handling materials, are likely to result in injuries.

PUSHING AND PULLING

Pushing and pulling are preferable to lifting and lowering objects; if done at proper height, rolling or sliding an object strains the body less. Use of pushing and pulling is particularly

---

Ergonomic Check 7-5. AVOID STRAINING MOVEMENTS

By proper design of workstation and by proper work practices

- avoid bending,
- avoid stooping, and
- avoid twisting.

---

preferable to lifting and lowering an object if it is heavy, fragile, or difficult to handle, and if handling must be done often.

Ergonomic Check 7-6.
ARRANGE FOR PUSHING AND PULLING

Arrange the work so that

- pushing (preferred) or
- pulling can be done

directly in front of the body, at about waist height.

To initiate the movement of an object, a larger force must be generated than is needed to sustain the movement. Both the initial push or pull, and the sustained effort, are facilitated if the object is put on a surface that allows easy gliding, or even better, rolling — see Key 3.

Ergonomic Check 7-7.
MAKE PUSHING AND PULLING EASY

Use equipment such as a roller conveyor or other equipment that makes pushing or pulling of an object easy.

Table 7-1 (adapted from Snook and Ciriello, 1991[34]) shows examples of acceptable push forces, while Table 7-2 shows examples of acceptable pull forces for American males and females.

[34] Snook, S.H. and Ciriello, V.M. (1991). The design of manual handling tasks: revised tables of maximum acceptable weights and forces. *Ergonomics*, 34, 1197–1213.

Table 7-1 Maximal Acceptable Push Forces (N)

	Height (a)	Percent (b)	One 2.1 meter push every							Height (a)	Percent (b)	One 30.5 meter push every				
			6 sec	12 sec	1 min	2 min	5 min	30 min	8 hr			1 min	2 min	5 min	30 min	8 hr
Initial forces																
Males	95	90	206	235	255	255	275	275	334	95	90	167	186	216	216	265
		75	275	304	334	324	353	353	432		75	206	235	275	275	343
		50	334	373	422	422	442	442	530		50	265	294	343	343	432
Females	89	90	137	147	167	177	196	306	216	89	90	118	137	147	157	177
		75	167	177	296	216	235	245	265		75	147	157	177	186	206
		50	196	216	245	255	285	294	314		50	177	196	206	226	255
Sustained forces																
Males	95	90	98	128	159	167	186	186	226	95	90	79	98	118	128	157
		75	137	177	216	216	245	255	304		75	108	128	157	177	206
		50	177	226	225	285	324	335	392		50	147	167	196	226	265
Females	89	90	59	69	88	88	98	108	128	89	90	49	59	59	69	88
		75	79	108	128	128	147	157	186		75	79	88	88	98	128
		50	168	147	177	177	196	206	255		50	98	118	118	128	167

(a) Vertical distance from floor to hands (cm); (b) Acceptable to 50, 75, or 90 percent of industrial workers. Conversion: 1 kg = 2.2 lb$_f$ = 9.81 N; 1 cm = 0.4 in.

Note: Use the complete tables by Snook, S.H. and Ciriello, V.M. (1991). The design of manual handling tasks: revised tables of maximum acceptable weights and forces. *Ergonomics,* 34, 1197–1213.

Table 7-2 Maximal Acceptable Pull Forces (N)

	Height (a)	Percent (b)	One 2.1 meter pull every						
			6 sec	12 sec	1 min	2 min	5 min	30 min	8 hr
Initial pull									
Males	95	90	186	216	245	245	265	265	314
		75	226	265	304	304	314	324	383
		50	275	314	353	353	383	383	461
Females	89	90	137	157	177	186	206	216	226
		75	157	186	206	216	245	255	265
		50	186	226	245	255	285	294	314
Sustained pull									
Males	95	90	98	128	157	167	186	196	235
		75	128	167	206	216	245	255	294
		50	157	206	255	265	304	314	363
Females	89	90	59	88	98	98	108	118	137
		75	79	118	128	128	147	157	196
		50	98	147	157	167	186	196	245

(a) Vertical distance from floor to hands (cm); (b) acceptable to 50, 75, or 90 percent of industrial workers. Conversion, 1 kg$_f$ = 2.2 lb$_f$ = 9.81 N; 1 cm = 0.4 in.

Note: Use the complete tables by Snook, S.H. and Ciriello, V.M. (1991). The design of manual handling tasks: revised tables of maximum acceptable weights and forces. *Ergonomics, 34,* 1197–1213.

CARRYING

Loads may be carried in many different ways. The appropriate manner depends on many variables: the weight of the load,

---

Ergonomic Check 7-8. GENERAL CARRYING RULES

- A load should be carried near the mid-axis of the body, near waist height. The farther away from the body's middle, either toward the feet or toward a side, the more demanding the carrying becomes.
- Carrying a medium load, 25 to 30 kg, distributed on chest and back is the least energy-consuming method.
- Carrying the load on the back, or well distributed across both shoulders and the neck (possibly with a yoke), also costs fairly little energy.

its shape and size, its rigidity or pliability, the provision of hand-holds or points of attachment, and the bulkiness or compactness of the load. Determining the best technique also depends on the distance the load is to be carried, on whether the path is straight or curved, flat or inclined, with or without obstacles, whether one can walk freely or must duck under obstacles or move around them. Table 7-3 lists different ways to carry loads and indicates some of their advantages and drawbacks.

Carrying a heavy load in one hand is especially fatiguing and stressful, particularly for the muscles of the hand, shoulder, and back. Nevertheless, it is often done because of the convenience of quickly grasping an object.

For carrying loads in the shop or factory, Snook and Ciriello (1991) published a table that indicates suitable weights to be carried by American males and females. This information is partially reprinted in Table 7-4.

Ergonomic Check 7-9. ROLL INSTEAD OF CARRYING

• Carrying may be replaced by rolling items.

For example, a simple platform cart, or conveyors of several kinds, are simple means to move objects more easily than carrying them.

If a load must also be picked up from a low location, or delivered at different heights, one may use equipment that not only rolls but also lifts and lowers the load.

LIFTING AND LOWERING

The need for lifting and lowering often indicates poor job design. In a shop or factory, the object should have been delivered to the work place at proper height and location — see Keys 2 and 3 — or lifted or lowered there by equipment to a height at which a human can push and pull it easily. Keep in mind that lifting or lowering any object also means raising and lowering body parts, often not only of hands and arms but also shoulders, head, neck, and torso — easily an added burden of 25 kg (more

Table 7-3 Techniques of Carrying Loads Totaling about 30 kg

Placement of load	Estimated energy expenditure on a straight flat path (kJ/min)	Estimated muscular fatigue	Local pressure and ischemia	Stability of the carrying person	Special aspects
In one hand	Very high (30; equal weights)	Very high	Very high	Very poor	Load easily manipulated and released
In both hands		High	High	Poor	Suitable for quick pickup and release; for short-term carriage even of heavy loads
Clasped between arms and trunk	Not measured	Unknown	Unknown	Unknown	Compromise between hand and trunk use
On head	Fairly low (22; when supported with one hand)	High if stabilized by hand	Unknown	Very poor	May free hand(s); strongly limits body mobility; determines posture; pad is needed
On neck	Medium (23; therpa-type strap around forehead)	Unknown	Unknown	Poor	May free hand(s); affects posture — If accustomed to this technique, suitable for heavy and bulky loads
On one shoulder	Not measured	High	Very high	Very poor	May free hands; strongly affects posture — Suitable for short-term transport of heavy and bulky loads
Across both shoulders	High (26; yoke held with one hand)	Unknown	High	Poor	May free hand(s); affects posture — Suitable for bulky and heavy loads; pads and means of attachment must be carefully provided

Location					
On back	Medium (22; for backpack High (25; bag held in place with hands)	Low	Unknown	Poor	Usually frees hands; forces forward trunk bend; skin cooling problem	Suitable for large loads and longtime carriage. Packaging must be done carefully, attachment means shall not generate areas of high pressure on body
On chest	Not measured	Low	Unknown	Poor	Frees hands, easy hand access; reduces trunk mobility; skin cooling problem	Very advantageous for small loads that must be accessible
On chest and back	20, lowest	Lowest	Unknown	Good	Frees hands; may reduce trunk mobility; skin cooling problem	Very advantageous for loads that can be divided/distributed; suitable for longtime carriage
At waist, on buttocks	Not measured	Unknown	Unknown	Very good	Frees hands; may reduce trunk mobility	Around waist for smaller items, distributed in pockets or by special attachments; superior surface of buttocks often used to partially support backpacks
On hip	Not measured	Low	Unknown	Very good	Frees hands; may affect mobility	Often used to prop large loads temporarily
On legs	Not measured	High	Unknown	Good	Easily reached with hands; may affect walking	Requires pockets in garments and/or special attachments
On foot	Highest	Highest	Unknown	Poor	Usually not useful	Occasionally used for exercising

Adapted from Kroemer, K.H.E., Kroemer, H.B., and Kroemer-Elbert, K.E. (1994). *Ergonomics: How to Design for Ease and Efficiency*. — especially Chapter 10. Englewood Cliffs, NJ: Prentice-Hall. With permission.

Table 7-4   Maximal Acceptable Carry Weights (kg)

	Height (a)	Percent (b)	One 2.1 meter carry every						
			6 sec	12 sec	1 min	2 min	5 min	30 min	8 hr
Males	79	90	13	17	21	21	23	26	31
		75	18	23	28	29	32	36	42
		50	23	30	37	37	41	46	54
Females	72	90	13	14	16	16	16	16	22
		75	15	17	18	18	19	19	25
		50	17	19	21	21	22	22	29

(a) Vertical distance from floor to hands (cm); (b) acceptable to 50, 75, or 90 percent of industrial workers. Conversion, 1 kg = 2.2 lb; 1 cm = 0.4 in.

Note: Use the complete tables by Snook, S.H. and Ciriello, V.M. (1991). The design of manual handling tasks: revised tables of maximum acceptable weights and forces. *Ergonomics,* 34, 1197–1213.

than 50 lbs). Yet, in many circumstances occasional lifting or lowering, and carrying, cannot be avoided. In these cases, the following guidelines apply:

- Lifting and lowering, if unavoidable, should be performed in front of the trunk, between hip and chest height.
- If lifting from the floor, or lowering to the floor, is absolutely necessary, it should be performed between the legs if at all possible. The major motion should be in the legs, not in the back (bend your knees).

Lifting or lowering objects in front of the feet and knees should be avoided because it requires excessive forward bending of the back, which is likely to be injurious.

To allow lifting or lowering in front of the trunk (preferred) or between the legs, the object must be of suitable size, compactness, and should be easily and securely grasped. Often, handles or cutouts for the hands are desired to allow secure coupling between hand and object — see Key 3.

The less often objects are lifted, the less strenuous the job and the lower the probability of fatigue and repetitive trauma injury. The less frequent the lifting, the larger the load that one can lift each time. (Yet, one should not draw the conclusion that therefore an occasional extremely heavy lift is permissible.)

Ergonomic Check 7-10.
GENERAL RULES FOR LIFTING AND LOWERING

Lift and lower

- in front of the trunk, and
- close to the trunk.

If the job requires other strenuous activities besides lifting, a person's metabolic, circulatory, respiratory, and musculoskeletal (central) capabilities may be exhausted, which makes the lifting even more strenuous. If it is absolutely necessary that heavy work including lifting must be done, suitable rest periods must be provided for the people involved. Rest breaks should be freely selected by the workers and encouraged by supervisors.

Ergonomic Check 7-11.
AVOID HEAVY WORK WITH LIFTING

- Do not require frequent lifting of heavy loads if the job is already strenuous by itself.

TEAM EFFORTS

If a load appears to be too heavy or too large to be handled alone, one should try to call another person to help. This person might help to stabilize and balance a load that is otherwise difficult to handle, but the strengths of two persons do not simply double. In team lifting, even under conditions of good coordination and suitable placement of hands and feet, usually only about 80% of the sum of lift strengths of two or three persons can be exerted. If the effort involves considerable load movement, usually only about two-thirds of the combined individual strengths can be expected. In large teams, especially when there is little motivation, one or more persons might not be able or willing to contribute fully, especially if the object is relatively small and difficult to grasp or maneuver.

Ergonomic Check 7-12.
INDIVIDUAL STRENGTHS DO NOT ADD

In team lifting and lowering,

- at best only about 80% of the sum of lift strengths of two or three persons can be exerted.

LIMITS FOR LIFTING AND LOWERING

Only a few decades ago, it was believed that certain set weights could be lifted safely by men or women or children. This simplistic idea is false for many reasons: one is that people are of different sizes, strengths, and skills; another is that the same load might be handled in many different ways; for example, according to Newton's Second Law, the force depends on the acceleration applied to the load, not simply on its weight (mass).

In 1981, the U.S. National Institute for Occupational Safety and Health (NIOSH) published its *Work Practices Guide for Manual Lifting*.[35] This document contained distinct recommendations for acceptable masses that must be lifted. These recommendations considered location and length of the lift path, and the frequency of lifting. In 1991, NIOSH[36] revised the guidelines and established a new Recommended Weight Limit (RWL). It represents the maximal mass of a load that may be lifted or lowered by 90% of U.S. industrial workers, male or female, who are physically fit and accustomed to physical labor.

The 1991 equation used to calculate the RWL resembles the 1981 formula (for the so-called Action Limit), but includes new multipliers to reflect the quality of hand–load coupling and the asymmetry of the load handling, meaning how far to the side the load

[35] NIOSH (Ed.) (1981). *Work Practices Guide for Manual Lifting*. DHHS (NIOSH) Publication 81-122. Washington, DC: U.S. Government Printing Office.

[36] Putz-Anderson, V. and Waters, T. (1991). Strategies for assessing multi-task lifting jobs. In *Proceedings of the Human Factors Society 35th Annual Meeting* (pp. 809–813). Santa Monica, CA: Human Factors Society; Waters, T.R., Putz-Anderson, V., Garg, A., and Fine L.J. (1993). Revised NIOSH equation for the design and evaluation of manual lifting tasks. *Ergonomics*. 36, 749–776; Waters, T.R. and Putz-Anderson, V. (1996). Revised NIOSH lifting equation. Chapter 31 (pp. 627–653) in A. Bhattacharya and J.D. McGlothlin (Eds.), *Occupational Ergonomics*. New York: Dekker.

is instead of directly in front of the body (feet). The 1991 equation allows as maximum a Load Constant (LC) (permissible only under the most favorable circumstances) with a value of 23 kg (51 lbs), which may not be exceeded under any circumstances. This is quite a reduction from the maximal 40 kg in the 1981 NIOSH guidelines.

Ergonomic Check 7-13. MAXIMAL LIFT LOAD

According to the 1991 NIOSH Guideline,

- at best only 23 kg (51 lb) may be safely lifted or lowered by a single person.

The following assumptions and limitations apply, among others:

- The equation does NOT include safety factors for such conditions as unexpectedly heavy loads, slips or falls, or for temperatures outside the range of 19°C (66°F) to 26°C (79°F) and for humidity outside 35 to 50%.
- The equation does NOT apply to one-handed tasks, while seated or kneeling, to tasks in a constrained work space, or to lifting people.
- The equation assumes that other manual handling activities and body motions requiring high energy expenditure such as in pushing, pulling, carrying, walking, climbing, or static efforts as in holding, are minimal.
- The equation assumes that the worker/floor surface coupling provides a coefficient of static friction of at least 0.4 between the shoe sole and the standing surface.

The equation may be applied under the following circumstances:

- The task is lifting or lowering, meaning manually grasping and moving an object of defined size without mechanical aid to a different height level.
- The time duration of lifting or lowering is normally between two and four seconds. The load is grasped with both hands.
- The motion is slow, smooth, and continuous.

- The posture is unrestricted (see above).
- The foot traction is adequate (see above).
- The temperature and humidity are moderate (see above).
- The horizontal distance between the two hands is not more than 65 cm (25 in).

For these conditions, NIOSH provides an equation for calculating the Recommended Weight Limit (RWL):

RWL = LC × HM × VM × DM × AM × FM × CM

LC   is the Load Constant of 23 kg (51 lbs).

Each multiplier can assume values between 0 and 1:

HM represents the Horizontal Multiplier, where H is the horizontal location (distance) of the hands from the midpoint between the ankles at the start and end points of the lift.

VM is the Vertical Multiplier, where V is the vertical location (height) of the hands above the floor at the start and end points of the lift.

DM is the Distance Multiplier, where D is the vertical travel distance from the start to the end points of the lift.

AM is the Asymmetry Multiplier, where A is the angle of asymmetry, i.e., the angular displacement of the load from the medial (midsagittal plane) which forces the operator to twist the body. It is measured at the start and end points of the lift, projected onto the floor.

FM   is the Frequency Multiplier, where F is the frequency rate of lifting, expressed in lifts per minutes. It depends on the duration of the lifting task.

CM   is the Coupling Multiplier, where C indicates the quality of coupling between hand and load.

To help apply the recommended weight limit, a Lifting Index (LI) is calculated: LI = L/RWL, with L the actual load. If LI is at or below 1 (one), no action must be taken. If LI exceeds 1, the job must be ergonomically redesigned.

The following values are entered in the equation for RWL:

	Metric	**U.S. customary**
Load Constant **LC**	23 kg	51 lbs
Horizontal Multiplier **HM**	25 cm/**H**	10 in/**H**
Vertical Multiplier **VM**	1 − (0.003|**V** − 75|)	1 − (0.0075|**V** − 30|)
Distance Multiplier **DM**	0.82 + (4.5/**D**)	0.82 + (1.8/**D**)
Asymmetry Multiplier **AM**	1 − (0.0032**A**)	
Frequency Multiplier **FM**	see box on next page for values of F	
Coupling Multiplier (**CM**)	see box on next page for values of C	

The variables H, V, D, and A can have the following values:

H *is between 25 cm (10 in.) and 63 cm (25 in.).* Although objects can be carried or held closer than 25 cm in front of the ankles, most objects that are closer cannot be lifted or lowered without encountering interference from the abdomen. Objects farther away than 63 cm (25 in.) cannot be reached and cannot be lifted or lowered without loss of body balance, particularly when the lift is asymmetrical and the operator is small.

V *is between 25 cm (10 in.) and (175-V) cm [(70-V) in.].* For a lifting task, D is equal to $V_{end} - V_{start}$; for a lowering task, D equals $V_{start} - V_{end}$.

A *is between 0 and 135 degrees.*

F *is between one lifting or lowering every 5 minutes (over a working time of 8 hours) to 15 lifts or lowers every minute (over a time of 1 hour, or less),* depending on the vertical location V of the object. Table 7-5 lists the Frequency Multipliers FM.

Table 7-5   Frequency Multipliers FM

Lifts/min	≤ 8 hrs		≤ 2 hrs		≤ 1 hrs	
	V <75	V ≥75	V <75	V ≥75	V <75	V ≥75
0.2	0.85	0.85	0.95	0.95	1.00	1.00
0.5	0.81	0.81	0.92	0.92	0.97	0.97
1	0.75	0.75	0.88	0.88	0.94	0.94
2	0.65	0.65	0.84	0.84	0.91	0.91
3	0.55	0.55	0.79	0.79	0.88	0.88
4	0.45	0.45	0.72	0.72	0.84	0.84
5	0.35	0.35	0.60	0.60	0.80	0.80
6	0.27	0.27	0.50	0.50	0.75	0.75
7	0.22	0.22	0.42	0.42	0.70	0.70
8	0.18	0.18	0.35	0.35	0.60	0.60
9	0	0.15	0.30	0.30	0.52	0.52
10	0	0.13	0.26	0.26	0.45	0.45
11	0	0	0	0.23	0.41	0.41
12	0	0	0	0.21	0.37	0.37
13	0	0	0	0	0	0.34
14	0	0	0	0	0	0.31
15	0	0	0	0	0	0.28
>15	0	0	0	0	0	0

Values for V are in cm.

C *is between 1.00 ("good") and 0.90 ("poor").* The effectiveness of the coupling may vary as the object is being lifted or lowered: a "good" coupling can quickly become "poor." Three categories are defined in detail in the NIOSH publication and result in the following values for the Coupling Multiplier CM:

Couplings	V < 75 cm (30 in.)	V ≤ 75 cm (30 in.)
Good	1.00	1.00
Fair	0.95	1.00
Poor	0.90	0.90

NIOSH personnel have provided examples on how to apply these guidelines to a variety of actual conditions at work.[37]

Snook and Ciriello in 1991 also published tables of maximally acceptable lift and lower forces, which where derived in psychophysical experiments (see Key 4). Excerpts are shown in Tables 7-6 and 7-7.

In addition to the information provided in 1991 by Snook and Ciriello and by NIOSH, excerpted above, other recommendations and guidelines have been published and/or are available as computer programs. Some of these agree fairly well with each other, at least in part and under certain conditions, but one may also find major discrepencies.[38] In Europe, for example, entirely different recommendations are applied. In the case of differing recommendations it is prudent to follow the set that protects the load handler best.

It is up to the ergonomist (engineer, manager, designer, health care provider) to determine the suitable conditions of material handling. As a rule, the job must be done, but not at the expense of the material handlers — see Figure 7-1 for how NOT to do it. The nature and details of the job are to be determined with the welfare of people foremost in mind. Therefore the ergonomist will select those conditions that are the easiest (least stressful, safest). Figure 7-2 depicts a general strategy.

---

[37] NIOSH (Ed.) (1981). *Work Practices Guide for Manual Lifting.* DHHS (NIOSH) Publication 81-122. Washington, DC: U.S. Government Printing Office; Putz-Anderson, V. and Waters, T. (1991). Revisions in NIOSH Guide to Manual Lifting. Paper presented April 9, 1991 at a conference at the University of Michigan; Waters, T.R. (1991). Strategies for assessing multi-task lifting jobs. In *Proceedings of the Human Factors Society 35th Annual Meeting* (pp. 809–813). Santa Monica, CA: Human Factors Society; Waters, T.R., Putz-Anderson, V., Garg, A., and Fine, L.J. (1993). Revised NIOSH equation for the design and evaluation of manual lifting tasks. *Ergonomics.* 36, 749–776; Waters, T.R. and Putz-Anderson, V. (1996). Revised NIOSH lifting equation. Chapter 31 (pp. 627–653) in A. Bhattacharya and J.D. McGlothlin (Eds.), *Occupational Ergonomics.* New York: Dekker.

[38] Hidalgo, J., Genaidy, A., Karwowski, W., Christensen, D., Huston, R., and Stambough, J. (1995). A cross-validation of the NIOSH limits for manual lifting. *Ergonomics.* 38, 2455–2462; Leamon, T.B. (1994). Research to reality: a critical review of the validity of various criteria for the prevention of occupationally induced low back pain disability. *Ergonomics.* 37, 1959–1974.

Table 7-6 Maximal Acceptable Lift Weights in kg

Width (a)	Distance (b)	Percent (c)	Floor level to knuckle height — One lift every								Knuckle height to shoulder height — One lift every								Overhead reach to shoulder height — One lift every							
			5 sec	9 sec	14 sec	1 min	2 min	5 min	30 min	8 hr	5 sec	9 sec	14 sec	1 min	2 min	5 min	30 min	8 hr	5 sec	9 sec	14 sec	1 min	2 min	5 min	30 min	8 hr
		Males																								
34	51	90	9	10	12	16	18	20	20	24	9	12	14	17	17	18	20	22	8	11	13	16	16	17	18	20
		75	12	58	18	23	26	28	29	34	12	16	18	22	23	23	26	29	11	14	17	21	21	22	24	26
		50	17	20	24	31	35	38	39	46	15	20	23	28	29	30	33	36	14	18	21	26	27	28	31	34
		Females																								
34	51	90	7	9	9	11	12	12	13	18	8	8	9	10	11	11	12	15	7	7	8	9	10	10	11	12
		75	9	11	12	14	15	15	16	22	9	10	11	12	13	13	14	17	8	8	9	11	11	11	12	14
		50	11	13	14	16	18	18	20	27	10	11	13	14	15	15	17	19	9	10	11	12	13	13	14	17

(a) Handles in front of the operator (cm); (b) vertical distance of lifting (cm); (c) acceptable to 50, 75, or 90 percent of industrial workers. Conversion, 1 kg = 2.2 lb; 1 cm = 0.4 in.

Note: Use the complete tables by Snook, S.H. and Ciriello, V.M. (1991). The design of manual handling tasks: revised tables of maximum acceptable weights and forces. *Ergonomics, 34,* 1197–1213.

Table 7-7 Maximal Acceptable Lower Weights in kg

Width (a)	Distance (b)	Percent (c)	Knuckle height to floor level One lower every								Shoulder height to knuckle height One lower every								Overhead reach to shoulder height One lower every							
			5 sec	9 sec	14 sec	1 min	2 min	5 min	30 min	8 hr	5 sec	9 sec	14 sec	1 min	2 min	5 min	30 min	8 hr	5 sec	9 sec	14 sec	1 min	2 min	5 min	30 min	8 hr
Males																										
34	51	90	10	13	14	17	20	22	22	29	11	13	15	17	20	20	20	24	9	10	12	14	16	16	16	20
		75	14	18	20	25	28	30	32	40	15	18	21	23	27	27	27	33	12	14	17	19	22	22	22	27
		50	19	24	26	33	37	40	42	53	20	23	27	30	35	35	35	43	16	19	22	24	28	28	28	35
Females																										
34	51	90	7	9	9	11	12	13	14	18	8	9	9	10	11	12	12	15	7	8	8	8	10	11	11	13
		75	9	11	11	13	15	16	17	22	9	11	11	12	14	15	15	19	8	9	10	10	12	13	13	16
		50	10	13	14	16	18	19	20	27	11	13	13	14	16	18	18	22	10	11	11	12	14	15	15	19

(a) Handles in front of the operator (cm); (b) vertical distance of lowering (cm); (c) acceptable to 50, 75, or 90 percent of industrial workers. Conversion, 1 kg = 2.2 lb; 1 cm = 0.4 in.

Note: Use the complete tables by Snook, S.H. and Ciriello, V.M. (1991). The design of manual handling tasks: revised tables of maximum acceptable weights and forces. *Ergonomics, 34,* 1197–1213.

Figure 7-1 Avoid back twisting and bending in material handling. Note that the elevated box reduces the back bending as compared to the box on the floor, but the worker must still lean forward to reach for the bottom. (Adapted from Kroemer, K.H.E., Kroemer, H.B., and Kroemer-Elbert, K.E. (1994). *Ergonomics: How to Design for Ease and Efficiency.* Englewood Cliffs, NJ: Prentice Hall. With permission.)

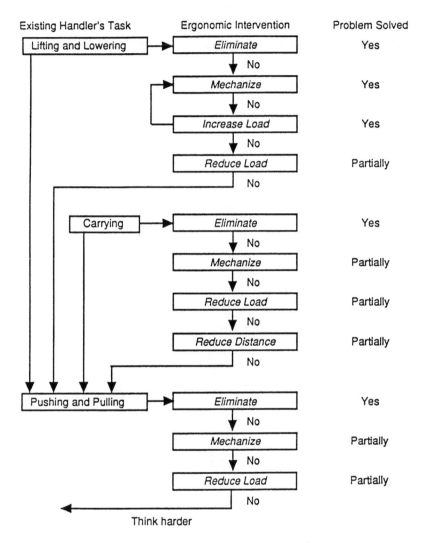

Figure 7-2 Ergonomic interventions in manual load handling.

CONCLUSION

Of course, the guidelines and recommendations discussed above do not absolutely ensure that a material handler will be satisfied with the work load, nor that he or she may not be overexerted or even injured when working. People and conditions differ from each other, from day to day. General guidelines and recommendations have many assumptions, clearly stated or often hidden in fine print. They simplify and normalize, for example, by assuming that only one type of work task is done (such as lifting) or that only one type of strain (such as spinal disc compression) loads the material handler. Furthermore, different experts and authorities (such as researchers, legislators, judges, government or international agencies) in various regions of the earth have differing opinions about what is tolerable or desirable in handling loads. What is acceptable in, say, Europe may not applicable in Asia or North America, for various reasons.

For up-to-date detail information, one has to follow the newest published literature. Of particular usefulness and importance are these fairly recent reports and summaries:

Ayoub, M. M. and Mital, A. (1989). *Manual Materials Handling*. London: Taylor & Francis.

Ciriello, V. M., Snook, S. H., and Hughes, G. J. (1993). Further studies of psychophysically determined maximum acceptable weights and forces. *Human Factors*, 35:1, 175–186.

Francis, R. L., McGinnis, L. F., and White, J. A. (1992). *Facility Layout and Location*. Englewood Cliffs, NJ: Prentice Hall.

Kroemer, K. H. E., Kroemer, H. J., and Kroemer-Elbert, K. E. (1997). *Engineering Physiology: Bases of Human Factors/Ergonomics* (3 rd ed.). New York: Van Nostrand Reinhold.

Kroemer, K. H. E., Kroemer, H. B., and Kroemer-Elbert, K. E. (1994). *Ergonomics: How to Design for Ease and Efficiency.* Englewood Cliffs, NJ: Prentice Hall.

Leamon T.B. (1994). Research to reality: a critical review of the validity of various criteria for the prevention of occupationally induced low back pain disability. *Ergonomics, 37,* 1959–1974.

Marras, W. S., Karwowski, W., Smith, J. L., and Pacholski, L. (Eds.) (1993). *The Ergonomics of Manual Work* (pp. 89–91). London: Taylor & Francis.

Mital, A., Nicholson, A. S., and Ayoub, M. M. (1993). *A Guide to Manual Materials Handling.* London: Taylor & Francis.

Randle, I. P. M., Nicholson, A. S., Buckle, P. W., and Stubbs, D. A. (1990). Limitations in the application of materials handling guidelines. Chapter 12 (pp. 90–97) in C. M. Haslegrave, J. R. Wilson, E. N. Corlett, and I. Manenica (Eds.). *Work Design in Practice.* London: Taylor & Francis.

Salas, E., Burgess, K. A., and Cannon-Bowers, J. A. (1992). Training effectiveness techniques. Chapter 14 in J. Weimer (Ed.), *Research Techniques in Human Engineering* (pp. 439–471). Englewood Cliffs, NJ: Prentice Hall.

Snook, S. H. and Ciriello, V. M. (1991). The design of manual handling tasks: revised tables of maximum acceptable weights and forces. *Ergonomics, 34*(9), 1197–1213.

Waters, T. R., Putz-Anderson, V., Garg, A., and Fine, L. J. (1993). Revised NIOSH equation for the design and evaluation of manual lifting tasks. *Ergonomics, 36*(7), 749–776.

In summary: Material handling is unproductive, costly, and hazardous. Therefore,

- reduce material handling to the absolute minimum; and/or,
- make material handling a mechanical process.

These are the *PRIMARY measures to take the hand out of material handling.*

If it cannot be avoided that people handle material, ergonomic considerations for the design of facility, job, equipment, and of the material to be handled become of utmost importance, because now "people are the kingpins of the work."

This results in *SECONDARY measures*:

- Fit the work system to the people.
- Use mechanical aids to eliminate hand labor.
- Design the work for a minimum of manual handling.

TERTIARY measures are:

- Assessment of existing physical job demands,
- Selection of individuals capable of doing the job, and
- Training of persons in proper material handling practices.

Elimination of all hazardous or overly strenuous load handling through proper work design is the strategy preferred over worker selection and training. An existing need for worker selection and training indicates that working conditions persist that are too strenuous and strainful.

ERGONOMIC design of material handling assures EASE and EFFICIENCY.

Index